John Winkler
11011 Cloverdale St.
Fredericksbrg, VA 22407

Comecon, trade and the West

Comecon, trade and the West

William V. Wallace and
Roger A. Clarke

St. Martin's Press, New York

© William V. Wallace and Roger A. Clarke, 1986

All rights reserved. For information, write:
Scholarly & Reference Division,
St. Martin's Press Inc., 175 Fifth Avenue, New York, NY 10010

First Published in the United States of America in 1986

Printed in Great Britain

Library of Congress Cataloging-in-Publication Data
Wallace, William V.
  Comecon, trade and the West.
  Bibliography: p.
  Includes Index.
  1. Council for Mutual Economic Assistance.
2. Europe, Eastern—Economic integration.   3. Communist
countries—Economic integration.   4. East–West trade
(1945–   )   I. Clarke, Roger A.   II. Title.
HC243.5.W35   1986   337.1′47   86-17837
ISBN 0-312-15104-7

Contents

List of Tables and Figure	vi
Introduction	vii
1 The origins of Comecon, and the barriers to co-ordination: Stalin to Brezhnev	1
2 The period of tension and drift: after Brezhnev	15
3 From drift to determination: towards Gorbachev	41
4 The Twenty-seventh Party Congress and the future of Comecon	55
5 Comecon and Common Market	67
6 The economic development of the CMEA member countries and the CMEA itself	78
7 The major economic problems facing the CMEA member countries in the 1970s and 1980s	115
8 Economic reforms and attempts to improve efficiency: prospects for the future	134
9 CMEA trade: problems, prospects and East–West relations	155
Conclusion	165
Select bibliography	170
Index	173

List of tables and figure

Tables

6.1	Annual Growth of National Income	107
6.2	Annual Growth of Gross Industrial Production	108
6.3	Annual Growth of Gross Agricultural Output	109
6.4	Annual Growth of Total Volume of Capital Investment	110
6.5	Population, Natural Increase and Growth of State Employment	111
6.6	Growth of Foreign Trade Turnover	112
6.7	Geographical Distribution of CMEA Countries' Foreign Trade	113
7.1	Structure of Employment, 1983	118
7.2	Energy and Steel Consumption per Unit of Gross Social Product, 1970	122
7.3	Changes in Steel Consumption per Unit of National Income	122
7.4	Energy and Steel Intensities	123
7.5	Structure of National Income Produced	129

Figure

5.1	Simplified Diagram of CMEA Decision-making Procedure	68

Introduction

One of the most important European and international developments since the Second World War has been the establishment and growth of the European Economic Community, or Common Market. It still has a long way to go to reach its full economic, not to mention its political, potential. But it is already a powerful force and, in media terms, a household word.

The same cannot be said about the Council for Mutual Economic Assistance, or Comecon, although it is almost a decade older, and although it is actually as well as potentially more powerful. Comecon is not a difficult word to say, but it does not roll readily off the tongue. This is equally true of its acronym, CMEA, which is bandied about much less than that of its opposite number, EEC.

That does not mean to say that the CMEA is unimportant. On the contrary. If since 1957 the Common Market has brought together in stages twelve countries from Western and Southern Europe, friends and former foes alike, Comecon has from its inception in 1949 embraced six East European countries—Poland, Czechoslovakia, Hungary, Romania, Bulgaria, and East Germany—and the enormous might of the Soviet Union. In 1962 it went on to admit Mongolia and, greatly extending its remit in 1972 and 1978, Cuba and Vietnam. It is territorially and geographically much more extensive than the Common Market, as well as richer in raw materials and rather more populous. Indeed, in one sense, the CMEA is a superpower beside which the Common Market pales into second or third rank.

Paradoxically, this is one reason why Comecon is not a household word. It includes the real superpower, the Soviet Union, and almost inevitably the title of the whole gives way to that of the dominant member. It has become more usual to talk about the Soviet Union or, at best, about the Soviet bloc. Since the Common Market does not include the United States, it has no such problem and goes by its own name.

Yet there is also a more subtle reason. The Council for Mutual Economic Assistance is more or less what it says it is. It is a committee, or series of committees, or an organisation; it is not a developing organism like the European Economic Community. It was Stalin who decreed it into existence and put 'mutual assistance' in its name—a description which must have seemed somewhat hypocritical to many East Europeans at the time. But the fact was that he did not wish to integrate Eastern Europe and the Soviet Union in the way that the EEC was meant to integrate Western Europe. And despite the efforts of his sucessors, CMEA shows few signs of integration either in achievement or in intention. So it is incorrect to speak of Comecon in the same way as we speak of the Common Market.

Equally, the emphasis within CMEA remains economic. Its function is to promote co-operation in industry, agriculture and trade, not in politics or in military affairs. Accordingly, since international issues are so often political or military, it is more correct to refer to the Soviet bloc or to the Warsaw Treaty Organisation or, in many cases, simply to the Soviet Union itself whose leadership, even supremacy in these matters is unquestioned. Admittedly, the EEC does not have an independent military existence but operates alongside the United States within the North Atlantic Treaty Organisation. Yet there are elements of autonomous strategic and tactical activities in the EEC; and it is attempting to develop its own foreign policy lines. For all these reasons, Comecon does not, and cannot, conjure up the same sense of importance as the Common Market.

Yet, just as with its growth and in face of an altering world economy the EEC is re-examining its role and its fortune, so, with the arrival of Gorbachev as General Secretary of the Communist Party of the Soviet Union, the CMEA has been undergoing a reappraisal. The difference in phraseology is important. The EEC re-examination is generated from several quarters, while the CMEA reappraisal has come from one side and from above. But in many ways the driving force is the same. Whether the CMEA is an adjunct to the Soviet Union or has a marginally autonomous existence, it has been performing economically no better than the Soviet Union. Equally, despite its military prowess and its success in space, the Soviet Union has been falling behind the United States in technology and productivity and runs the risk of becoming a second class superpower as well as a socialist society with inferior social conditions. As the representative of a new generation— though not the newest—Gorbachev has risen to power to arrest this comparative decline. And the CMEA is one of the objectives, as well as one of the potential instruments, of the economic regeneration that he proposes.

Previous General Secretaries made somewhat similar attempts to breathe new life into CMEA by trying to substitute deliberate integration for more or less accidental and essentially half-hearted co-operation. Khrushchev failed in the 1960s; Brezhnev had only slightly more success in the 1970s. The fault was not entirely theirs. The asymmetrical and centrifugal system they inherited from Stalin put both too much responsibility and too little power in their hands, while within the Soviet Union itself they lacked either the ideas or the machinery, or both, to make its economy independently successful.

Gorbachev has set out to reform and reinvigorate both the Soviet Union and the CMEA. He faces similar problems, though he may have a more consistent or a more perceptive grasp of what needs to be done; and the urgency of a response to the international challenge from the EEC itself, and from Japan and China, as well as from the United States, may be much greater and therefore more effective. His main concern is obviously with the Soviet Union itself, but his prospects in that respect are only derivatively the subject of this book. His prospects with the CMEA—and indeed his performance in his first year of office—are very much its subject. Interest lies not merely in the fortunes of Gorbachev, though he has succeeded in attracting more respect than Khrushchev and more liking than Brezhnev. The fate of the CMEA has importance for the prosperity and stability—or otherwise—of all its members, including the USSR, for an expansion or contraction in world trade—including that in which the West participates—and for a decrease or increase in international tension generally.

The barriers to integration that have built up within CMEA over the years are indeed formidable. Even co-ordination is suspect and often ineffective. On the other hand, there are pressures that the Soviet Union can exert, and there are attractions that have overcome East European doubts and disinclinations in the past. Until recently, Soviet energy supplies, for example, were both pressure and attraction. Gorbachev appears to have opted for co-operation, tending towards co-ordination. But building on such successes as Brezhnev was able to register, he has put the main emphasis on specific joint activities intended to utilise CMEA resources in order to begin catching up seriously on Western levels of productivity and technology by the year 2000. This was the clear message in his address to the Twenty-seventh Congress of the Soviet Communist Party in February 1986 in which he lauded the CMEA's so-called comprehensive programme of scientific and technological progress. And in spite of a certain lack of sparkle at subsequent party congresses in Czechoslovakia, Bulgaria and East Germany, and even in Poland, the notion is one that could find a sympathetic echo in Eastern Europe, particularly if it begins to look like having a chance of success. Other CMEA members may be pessimistic, even resistant; but they mostly seem willing to give Gorbachev his opportunity to justify some optimism—which may be asking rather a lot in the circumstances of the Soviet Union itself.

It is of course in their interest that Gorbachev should succeed. The alternative of continuing economic malaise, coupled with bouts of political bullying or military intervention of the kind the Soviet Union has indulged in in the past, is not one they would like to contemplate. In addition, the Gorbachev formula postulates reasonable trade with the West, reasonable in the double sense of avoiding a further accumulation of debt and at the same time securing high-technology imports. Some East European states, most notably Hungary, have long advocated an expansion of trade with the West, based on full convertibility. But an equally strong, or even more powerful, argument has frequently been deployed, particularly in the Soviet Union, in favour of its reduction to the point almost of extinction. So for the moment, at least, Gorbachev has saved other CMEA members from having to make what limited economic—and political—progress they could within an isolated and exclusive trading area.

The CMEA approach to the EEC in the summer of 1985 can be read both positively and negatively. Recent indications from Moscow suggest that Gorbachev's intention is in fact to try to increase the flow of goods between the two groups. However much this might be of assistance to the CMEA collectively as well as to most of its members individually, it may not be at all easy to achieve. The fall in the world price of oil is only one of the possible obstacles. Another is the difference in conception and in organisation of the two groups, which is discussed in Chapter 5. There is also the quite basic problem of what to trade when West European industry mostly outclasses East European, and when West European agriculture is already excessively productive.

This problem arises even more acutely in terms of trade with the United States and Japan. Trade with the developing countries is rather easier but mostly fails to produce a return in high technology. Unfortunately, but inevitably, the difficulty to be overcome makes for difficulty in overcoming it. The CMEA economies, and the CMEA economy as a whole, need better management and greater enterprise, extra investment and improved technology, in order to trade successfully with the West in general; but to satisfy this need, they first have to trade successfully with the West. This is the paradox discussed in the chapters that deal with economic problems and prospects. It is one that will not easily be resolved. At the heart of it is vested opposition to systemic reform, an ideological and individual obstacle as old as the first communist state. It remains to be seen whether Gorbachev can go beyond declarations of intent to practical changes that will create a sufficiently new system.

Close to the heart of the paradox, however, is the attitude of the West. The case is frequently argued that successful capitalism has no obligation to underwrite ailing socialism. From time to time an even more hostile view is presented. However, a majority in the West probably accepts that the world is in great need of a reduction in tension with the East, and that an expansion in two-way trade could act as a component in this process. But the argument can be developed in a different way. Capitalism—if there is such a thing—is not so successful these days that it can reject a possible increase in its trade or turn its back on its own philosophy of the international division of labour. That there can be a modest growth in the long run is another theme of this book; and that such a development would be beneficial to CMEA and so the West as well is perhaps implicit throughout.

Chapters 1–5, which offer a general overview, are the responsibility of Professor Wallace, Director of the Institute of Soviet and East European Studies in the University of Glasgow, and Chapters 6–9, which provide an economic analysis, are the responsibility of Roger Clarke, Lecturer in the Institute and Editor of its journal *Soviet Studies*.

1 The origins of Comecon, and the barriers to co-ordination: Stalin to Brezhnev

Because Comecon, with its massive headquarters in Moscow, is now an apparently active agency trying to co-ordinate the economies of its ten member states with their more than 450 million people, there is a tendency to regard it, like the Common Market, as an institution dedicated to that purpose from the start. But this is far from the case.[1]

The impact of the early years, to 1956

There is still some uncertainty about the precise origins of Comecon in 1949; so much of the Soviet evidence from that period is lacking. But Stalin's reasons for establishing it appear to have been more negative than positive, as was the case with much of his policy towards Eastern Europe at the time. Following what he perceived to be an American drive for control of the whole of Europe, outside the Soviet Union, particularly through the establishment of the Marshall Plan in 1947, he clearly wanted to establish a distinct and defensible Soviet–East European economic system. He was anxious to keep other powers out of neighbouring buffer states rather than to integrate them into a new and mammoth economy. This is borne out by the foundation documents, which speak of co-operation, not of co-ordination, still less of integration, and which, despite a year of intense discussion, were not elaborated.

Much changed in the Soviet Union after Stalin's time. Yet this negative attitude persisted, or at least regularly reappeared. Moscow's behaviour towards Czechoslovakia in the late 1960s was conditioned in part by a fear that it would gradually slip into the Western economic system through extended, market-determined trade; and its view of the emergence of Solidarity in Poland in 1980 was that it was proof of the destabilising and disintegrating effect of dependence on Western investment.

On the other hand, Stalin seems also to have envisaged Comecon as the economic counterpart of the Cominform, which he had instituted in

1947 as the cover for the export of communist political forms, particularly to Eastern Europe. If it was necessary or opportune to extend Soviet power outwards, then it should be done on a political and economic basis that would reinforce the Soviet view of socialism both at home and abroad. Stalin negotiated bilateral economic treaties with each of the East European states that he reduced to the role of satellite and encouraged all of them to follow the path of autarchic industrialisation and collectivisation along which he had driven the Soviet Union. In this way he imposed control and ensured support. Far from encouraging integration, in practice he prevented co-operation. He extended the bilateral principle to relations among the East European states. He had earlier vetoed emerging customs unions; and he now ensured that the only economic connections among the East European states were bilateral agreements for a specific limited purpose such as the exchange of surplus industrial products.

In this respect, too, much changed after Stalin's time. Yet bilateralism persisted as a standard Soviet approach, as did the tradition of insisting on economic models that would not challenge accepted Soviet practice and the communist hierarchy that it supported. The fear of the power that would slip away from the Party in Eastern Europe and even in the Soviet Union itself lay behind Moscow's refusal to allow the idea of a market to develop as far as the Hungarians, for example, would have liked.

The communists who came to power in Eastern Europe after the Second World War generally welcomed the Soviet exclusion of a Western economic presence that seemed previously to have done neither them nor their countries anything but harm. Those who survived into the early 1950s also accepted the imposition of the Soviet economic pattern as a genuine way out of past backwardness and dependence. And in some ways it was, as rising indices of industrial and agricultural production showed.

Yet once the first burst of extensive development was past, in the mid-1950s, the East European economies faltered and certainly failed to meet widespread public expectations. The result was a degree of dissatisfaction with the Soviet economic model and resentment at the way it had all too obviously been imposed from outside. Khrushchev's revelations in 1956 about Stalin's behaviour towards his East European neighbours, among others, merely exacerbated these feelings of distrust and dislike by confirming suspicions about the heavy price Stalin had exacted for the supposed benefits of a close trading relationship with the Soviet Union. And although subsequent Soviet leaders were in some ways an improvement, the East European view of the advantages of Comecon remained rather sceptical ever after.

The bringing together of the Soviet and East European economies would have been a difficult exercise in any case. The disparity in resources was enormous; the Soviet Union possessed, for example, 90 per cent of the land and energy and 70 per cent of the population. There were similar disparities among the East European countries themselves—Czechoslovakia, for instance, being highly industrialised by contemporary standards and Bulgaria hardly at all. There was, therefore, an initial tendency for each member of Comecon, including the Soviet Union, to scrutinise carefully the possible advantages and disadvantages to itself of the proposed co-operation. There was also no tradition of working together. Before the Second World War the East European countries and the Soviet Union had for the most part been on opposite sides of an economic fence, and certainly since the Depression Eastern Europe itself had been divided by autarchy.

Strangely enough, Stalin's policy tended to reinforce this separateness. The economic revolution he exported was the model of socialism in one country. All the East European states were incited to copy the Soviet Union in building up their own comprehensive and self-sufficient industry and agriculture, a pressure Stalin increased to meet the industrial needs of the Korean War between 1950 and 1953. Despite growing trade with the Soviet Union, they were effectively encouraged not to co-operate with one another or to contemplate a division of labour even with the Soviet Union. They all tended to develop the same strengths and weaknesses. This reinforcement of self-interest had long-lasting consequences as, for instance, when Romania fought tooth and nail against the specialisation that Khrushchev began to advocate in the early 1960s. Romania was determined to have a metallurgical industry of its own, irrespective of its lack of ore and the availability of iron and steel from other CMEA countries. It was equally determined not to be assigned to what it regarded as the lesser and the less profitable role of supplier of agricultural products for its neighbours.

Detailed central planning was another characteristic of the Stalinist economy that worked against the co-operation theoretically inherent in membership of Comecon. It gave enormous power to the planning bureaucracies and to the coterie of interests that made up the East European governments. They not only enjoyed immense powers of economic decision; they acquired an unwillingness to give them up to any extra-territorial authority. The fact that they possessed rather fewer powers of political decision made them all the more jealous of the economic rights they could exercise. So, from the start, East European governments were natural enemies of any co-operation that would be self-weakening. And it was on this rock that many schemes for strengthening the machinery of Comecon foundered, notably those propounded by Khrushchev.

Willingness to accept some form of external, supranational control of its economy also characterised the Soviet attitude to Comecon from the start. It was no part of Stalin's intention to construct a regional authority that would enable his East European comrades to gang up together and force a majority policy on him. It has even been suggested that he devised weak machinery for Comecon to keep the East European states in disagreement with one another and powerless against the Soviet Union. Whatever the truth of that, the machinery that emerged in 1949, although dedicated to promoting co-operation, seemed designed to prevent it, or at least to retard it. The central body, the Council Session, had to reach unanimous agreement in order to respect the sovereign rights of its member states. It could not make decisions, only recommendations. These had to be ratified subsequently by all governments, and even if so ratified, still had to be translated into policy by all governments acting separately. No independent permanent secretariat was established; and no real power was devolved upon subsidiary committees. After 1949, of course, the machinery was elaborated, and elements of it acquired greater power. But its basic weakness remained a serious obstacle to closer integration as well as a monument to the fears and antipathies of those who established it.

In some ways it was curious that Stalin did not design an organisation that would embody and perpetuate Soviet predominance. Yet to a considerable extent, Comecon was a sham. Stalin ruled Eastern Europe, as he ruled the Soviet Union, by other than constitutional machinery. Subsequently, Khrushchev attempted to remove some of the sham and relaxed the tight Soviet political grip on the East European states. But the damage was done. Indeed, much of the later East European attitude to Comecon had little to do with economics at all. What was remembered was the terror that Stalin conducted in the last few years of his life, that swept many leading East European communists to their deaths and condemned many innocents, communists and non-communists alike, to fierce terms of imprisonment, and that manipulated the survivors to serve purely selfish Soviet ends at home and abroad. The Cominform did not long survive Stalin, and its dissolution was not mourned. Comecon did survive, but it ever after had to live down the legacy of Stalin's callous regimentation of Eastern Europe.

What was also remembered from the Stalin period was that Yugoslavia was expelled from the Cominform in 1948 and not admitted to Comecon in 1949 and yet survived politically and prospered economically. In the immediate post-Stalin period, too, it continued to improve its position and, in 1964, it even managed to negotiate the special status of 'limited participant' in the activities of Comecon. This was in sharp contrast to the post-1953 political experience of other East European states. In the wake of Khrushchev's famous anti-Stalin speech to the Twentieth Congress of the Soviet Communist Party in 1956, Poland managed to make most headway, gaining for itself a special relationship with the Soviet Union which allowed it a degree of autonomy in its internal political and economic policies in return for solid general support of the Soviet line internationally and throughout Eastern Europe. Hungary, by contrast, stumbled into pressing excessive demands and suffered the full weight of Soviet military repression, a terrible lesson to the rest of Eastern Europe in the unlikelihood of fundamental change. Even Czechoslovakia, which had been showing signs of reasserting its own view of socialism, crept back into its realistic shell, while the rest of Eastern Europe scarcely murmured. By the end of 1956 a political situation inimical to the development of genuine economic co-operation through Comecon seemed simply to have been confirmed.

The impact of the later years, 1956–80

The years 1949–56, which represent the origins of Comecon, are now more than thirty years away. There has been ample time since for further legacies and memories to accumulate, affecting both the development of Comecon and attitudes towards it. And these, too, have helped to shape both current problems and future prospects.[2]

One ineradicable impact of 1956 was on Hungary itself. There was never afterwards any serious suggestion of Hungary embarking alone on a course of major political reform. Quite the reverse. Politics was eschewed, and the Hungarians concentrated their very considerable talents on modernising their economy. For a while this was true throughout Eastern Europe. The veto on independent political initiatives drove governments to experiment in other directions; and the need to divert popular disquiet afer the shooting down of workers in

Budapest impelled communist parties to seek ways of rapidly improving living standards. In the Soviet Union, Khrushchev was hoist with his own petard; having risen to power on his criticism of Stalin, he had to concentrate on attempting to succeed where Stalin had failed, in the diversification of industry and the raising of agricultural output. Throughout the Soviet bloc, therefore, the emphasis was on the individual national economies, and little immediate thought was given to possible co-operative efforts. In any case, in the aftermath of Budapest, the idea of co-operation had rather a hollow sound.

On the other hand, Khrushchev was anxious to undo some of the damage of 1956 and to find a new political and economic relationship with Eastern Europe. He also discovered himself to be in a steadily changing international situation. Emerging friction with China demanded some kind of political *modus vivendi* with Eastern Europe; at the same time peaceful competition with the West, a policy dictated by the impossibility of contemplating a nuclear war, postulated a marked improvement in the overall economic performance of the Soviet Union and Eastern Europe, all the more so as the signs grew clearer of the likely emergence of a West European Common Market. The Moscow Declaration of 1957 ostensibly put the relationship between the Soviet Union and other socialist countries on the basis of equality, respect and non-interference. Talks began on improving the machinery of Comecon and on developing the division of labour. The years 1959–60 saw the emergence and acceptance of the first reasonably detailed Comecon Charter, and 1962 the Basic Principles of the International Socialist Division of Labour.

However, although these two documents marked a genuine advance in thinking about working together, and in committing thoughts to paper, they also enshrined some of the existing barriers to co-operation and helped to create fresh ones. References in the Charter to the member states' co-ordinating their efforts were accompanied by others that were partly contradictory, concerning equality, respect and non-interference. In addition, once Romania in particular realised that the Basic Principles, with their advocacy of specialisation according to comparative advantage, would place it in a second- or third-class category as a supplier of raw materials rather than in the first rank as an economic all-rounder, the notion of a division of labour came under attack. What was then noticed, too, was that the Moscow Declaration of 1957 had also referred to the principle of fraternal mutual aid governing the relations among socialist countries, a principle that could be used to justify another Budapest. At the same time, China stirring the pot as part of its now bitter quarrel with the Soviet Union did nothing to encourage the Romanians to risk their remaining sovereignty by falling in with Khrushchev's ideas on Comecon. And inherent in Khrushchev's concept was developing CMEA in such a way as to make it economically superior to the capitalist world, irrespective of the national consequences.

This new appreciation of the possible consequences of closer co-operation reinforced another emerging tendency in the Soviet Union and Eastern Europe, the tendency independently to seek new or amended forms of internal economic organisation. Khrushchev's initial bold but curious attempt to bring Soviet industrial and agricultural performance nearer to that of Western countries failed conspicuously. In the meantime, most of the East European economies had reached the limits of extensive development. How to switch them successfully in the direction of intensive development—and how to increase productivity in the Soviet Union—became urgent matters of individual concern, turning attention away from the long-term potential of Comecon to the more immediate question of local economic reform. The beckoning prosperity of the EEC also appeared to offer more attractive bilateral opportunities to Comecon countries separately than any Comecon multilateral scheme; and peaceful coexistence, or *détente* as it came eventually to be termed after Brezhnev succeeded Khrushchev in 1964, raised no political obstacles.

Brezhnev's initial preoccupation, indeed, was with undoing some of the more conspicuous of Khrushchev's domestic failures. In September 1965 Kosygin introduced a series of economic reforms giving enterprises a comparatively greater degree of decision-making power and raising the status of profit as a criterion of economic efficiency and a stimulus to it. There even followed a debate on the role of market and prices. Thought was given to the import of Western technology, and a generally more commercial approach was adopted to trade with the Third World. It was therefore possible for other Comecon members to proceed with plans for reform; and under Soviet guidance the Council agreed in November 1967 to introduce monetary measures to allow inter-enterprise trading. Hungary edged steadily towards the New Economic Mechanism that it finally established in 1968, and that took it furthest away of all the East European systems from the highly centralised Soviet model and raised yet another obstacle to closer co-operation. Minor decentralising reforms were introduced in Poland, East Germany and Bulgaria. Czechoslovakia used the same interval to discuss economic changes necessitated by the apparent exhaustion of what had once been Eastern Europe's most advanced industrial country. An obscurantist political leadership procrastinated on improvements until its removal early in 1968 precipitated a rush towards economic and political changes. But by the middle of that year, opinion at the top in the Soviet Union was beginning to turn against reform and in particular against money mechanisms. In such an atmosphere the extent of the changes in Czechoslovakia was deemed excessive. And the invasion of Czechoslovakia in August put an abrupt end to widespread reform throughout Eastern Europe.

This was an important moment for the further development of Comecon. In the negative sense, the Soviet Union reiterated forcefully the limits of its political and economic tolerance. Within its bloc nothing could be allowed that would threaten, as seemed likely to happen in Czechoslovakia, the leading role or the state-wide unity of a communist party; nor could a market be permitted to assert itself, either within a single country like Czechoslovakia or still less in Comecon as a whole, to the point of endangering central planning or of encouraging a closer relationship with the West than with the East. This was a stance that also had the backing of Poland, East Germany and Bulgaria, though only the passive acquiescence of Hungary (and not even that from Romania). The so-called Brezhnev Doctrine that quickly spelt out the new relationship among socialist countries significantly watered down the earlier Moscow Declaration's references to sovereignty and put most of its emphasis on the responsibility of communist parties to socialism both in their own countries and throughout the 'Socialist Commonwealth'.

In the positive sense, the Soviet Union followed the invasion of Czechoslovakia with various attempts to reassert its sole and pre-eminent leadership of the world communist movement, such as through the not particularly successful World Communist Conference in 1969. It also embarked on a serious campaign to breathe life into Comecon, partly as an aid to its own much-needed economic development in the face of new international challenges from the United States and China, but even more as a means of preventing the recurrence anywhere in Eastern Europe of a Czechoslovak-like challenge. The now discredited principles of the international socialist division of labour were abandoned in favour of the idea of socialist economic integration; and this took documentary form in 1971 in the so-called Complex Programme which still serves as a basic guide to Comecon's objectives.

Despite Polish and especially Hungarian advocacy, the Complex

Programme gave little ground to the idea of a market-based organisation. But it went well beyond previous statements of intent in envisaging not merely co-operation among the various economies, but full-scale integration. It comprised seventeen different reports drawn up by Comecon agencies and dealing with diverse aspects of Comecon activity. It was a major Soviet effort. It also had the powerful support of East Germany, then feeling exposed in the midst of East–West negotiations for a settlement of the German position, and the loyal and dependent backing of Bulgaria. The new Czechoslovak regime was almost more royalist than the king, though the attitude of Poland was somewhat ambivalent as it tried to reap the economic advantages of integration without the accompanying challenge to its admittedly limited autonomy. Hungary was reluctant, anxious to protect its New Economic Mechanism and still convinced of the need for a system based on market prices; but it could not resist. The Complex Programme could therefore reach out much further than any previous plan. It postulated the joint construction of special projects and laid down rules for co-operation in specific areas such as engineering, mining and transportation. It also established plan co-ordination as a major Comecon aim and initiated discussions on appropriate improvements in Comecon machinery.

For all that, it was not a major breakthrough. There remained contradictions in the language of the Programme that mirrored differences of view among Comecon members. Partly because of the Romanian attitude, recognition was given to the wish of individual countries, despite growing integration, to maintain locally balanced economies. To prevent any one member holding the others back, all were given the right to declare their 'lack of interest' in a particular proposition so that it could still go ahead without them. Yet, subsequently, Romania was not the only one to neglect to declare its lack of interest and so to maintain a veto. The proposed new machinery also remained international in concept and, despite its declared purpose, was not envisaged as in any way supranational, at least not at that stage. In any case, much of the programme was aspiration, not achievement.

In the next few years there were successes that should not be underrated. A number of these came together in the so-called Concerted Plan of Multilateral Integration Measures of 1974–5, which finalised the details of several joint ventures in the exploitation of energy and raw materials, particularly on the territory of the Soviet Union, and instituted various schemes of scientific and technological collaboration. Long-term Target Programmes of Co-operation, covering energy and raw materials, engineering and agricultural products, consumer goods and transportation, began emerging in 1975–6 and are still being processed. Yet, apart from the difficulties these ran into, they still represented a piecemeal approach to integration. Institutional improvements, too, materialised mainly for specific areas of cooperation; and machinery for plan co-ordination remained elusive.

A major obstacle was the historical one. An institution designed for an earlier purpose could not easily be adapted for a new one, despite a fresh burst of energy. In establishing the European Economic Community in 1957, the original six West European countries practically built from scratch. In addition, no one of them had just demonstrated its immense superiority over the others, as the Soviet Union had by its invasion of Czechoslovakia. Distrust of Russia died hard; jealous of remaining individual rights persisted. The founders of the EEC also shared a common economic philosophy, whereas the Complex Programme papered over the differences between the Soviet preference for out-and-out planning and the Hungarian hankering after convertible currencies. During the 1970s, too, the Soviet Union had internal preoccupations. It also developed *détente* through the final peace settlement with Germany and the SALT talks with the United States, and it could hardly object to a few East European countries then interesting themselves more in the West. To some extent it even led the way towards solving the socialist countries' economic problems by importing Western technology through trade and, rather more, through Western loans. The world oil crisis that first emerged in 1973, and the gradual awareness of a raw materials shortage, pushed Comecon towards specific areas of collaboration. But otherwise the atmosphere of the mid- and late-1970s reverted to that of reluctant partnership.

The integration drive of the 1970s came from the Soviet Union. Perhaps inevitably, the headquarters of Comecon itself and many of its committees were in Moscow. This did not necessarily mean that the organisation was purely and simply a Soviet agency. On the other hand, Brezhnev did begin fairly early on to use Comecon, if not as an instrument, at least as an attraction in Soviet foreign policy, or in that part of it which is termed socialist foreign policy. Mongolia had already been admitted to membership in 1962 on a basis akin to that of Russia's western neighbours. However, in 1972 Brezhnev admitted Cuba, and in 1978 he went on to introduce Vietnam. In between times he also accorded observer status to a number of scattered African and Asian countries, all as states committed to socialism. The professed objectives were to extend specific aid following the ravages of imperialism and then to accelerate general economic development. The benefits were not expected to be one-sided, since European Comecon stood in need of certain raw materials and agricultural products. Yet the East European members showed less enthusiasm than the Soviet Union, apparently suspecting that they would have to bear a considerable part of the cost of levelling up the less developed economies of their new friends. And so a new factor was added to those already retarding integration, quite apart from the problem of trying to get more members to agree where fewer had failed.

The doubts that regularly floated to the surface during the 1970s were not, of course, exclusively East European. Particularly once the Soviet Union came to appreciate how heavily it was subsidising the East European economies by selling them oil after 1973 at below world world price, it began to to show signs of concern; it first began to charge more in 1975. Ironically, when it did so, it introduced a system based on world market prices, which ran counter to its own professed beliefs, and which threatened Eastern Europe with an eventual rise in energy costs it could barely afford. On a much broader political front Brezhnev also followed his Czechoslovak action with an intensification of top-level exchanges with his East European allies. He created a Political Consultative Committee and a Foreign Policy Committee of the Warsaw Pact and held regular meetings with Party and government leaders. But if he was concerned to carry his East European colleagues with him on a whole range of political and economic issues, there were signs that some of them resented the pressures thus applied to them. Nor were these mutual resentments confined to the leaders of the Soviet Union and Eastern Europe. In the late 1970s there was a steady growth in popular antipathies, the clearest example of which was in the Polish crisis.

The Polish crisis

The collapse of the Gierek regime in 1980 was the direct result of combined economic and political failure. Western loans and technology had been misused. Instead of rising rapidly as promised, the standard of living was falling drastically. Gierek's only answers were further promises and time-worn repression. Sixty per cent of the population was under thirty years of age, impatient and unimpressed, and sickened by emerging examples of corruption. Solidarity became the vehicle for popular unrest. Discontent, however, was not directed against the Soviet Union. It was a purely Polish affair.[3]

Or so it was at the start. But as the Polish Communist Party disintegrated, the whole paraphernalia of Soviet-type rule came under criticism; and it began to appear as if the growing vacuum would be filled by a political trade union. This was totally unacceptable to the Soviet Union. Since 1956 it had allowed Poland somewhat greater freedom to determine the details of its internal economic and political functioning on the understanding that there would be no challenge to basic Soviet interests. From Moscow's point of view Poland had already become too dependent upon the West financially. To permit the state now to lose control of industry as it earlier had of agriculture would be to accelerate the swing from a centrally-planned to a market economy with all the attendant dangers to the position of the Communist Party in Poland and of Poland to the Soviet bloc. Further, to allow a freely elected trade union to displace the Communist Party from its leading role in politics would be to run the risk of creating, at best, another Yugoslavia in the heart of Comecon and the Warsaw pact, and possibly something rather worse.

Under Brezhnev the Soviet Union did not intervene openly in the Polish crisis the way it had in Hungary in 1956 or in Czechoslovakia in 1968; but it clearly exerted considerable pressure both before, during and after the military takeover. The fact that it did not act overtly may have stemmed perhaps from the success of its indirect pressure. On the other hand, the fact that it was so keen to secure an internal Polish solution—to the extent of letting the Polish Communist Party go into limbo and then be recreated very gradually—was evidence, among other things, of its acute awareness of Polish anti-Russian feelings. For, whatever the origins of the Polish crisis, the background against which it was played out and Soviet reactions to it clearly produced the biggest outburst of anti-Russian nationalist feelings in Eastern Europe since the war. Perhaps Yugoslav feelings following 1948 were similar, though there was then a strong element of straightforward antagonism between opposing dictators. In Hungary in 1956, however, it could be argued that much of the active resentment was centred in Budapest; and in Czechoslovakia in 1968, that much of it was intellectual. But the Polish revolt of 1980 spread widely in 1981 among workers, peasants and people of all walks of life and of every age group; and backed by centuries of Polish tradition and generations of Soviet experience, it took on an increasingly anti-Russian flavour. Lastly, it was a revolt that had the support of an alternative and well-established ideology, the

Catholic Church, and therefore could not lightly be brushed aside. Solidarity itself might have been rendered politically largely impotent; but the feelings behind it remained a serious challenge. In this situation the Soviet Union apparently recognized that, apart from other internal and external pressure on it to proceed with caution, it had to accept that popular feelings were now a much more important factor than before in its dealings with Eastern Europe. Add to that the public reactions to Polish behaviour evident in the Soviet Union and several East European countries and it is possible to conclude that the centrifugal characteristics of the area had been strengthened, with all such a development implied for the future of Comecon.

Notes

1. For an introduction to Comecon and the East European economies see J. M. P. van Brabant, *Socialist Economic Integration: Aspects of Contemporary Economic Problems in Eastern Europe*, Cambridge, Cambridge University Press, 1980; and A. H. Smith, *The Planned Economics of Eastern Europe*, London, Croom Helm, 1983. See also Ye. T. Usenko, *The Multilateral Economic Cooperation of Socialist States*, Moscow, Progress Publishers, 1977; and O. T. Bogomolov, *Strany sotsialisma v mezhdunarodnom razdelenii truda*, Moscow, Nauka, 1986.
2. The literature on this period is now extensive. The following is a selection:

    On the general economic situation: S. Ausch, *Theory and Practice of CMEA Cooperation*, Budapest, Akadémiai kiadó, 1972; B. N. Ladygin, O. K. Rybakov and V. I. Sedov, *The Socialist Community at a New Stage*, Moscow, Progress Publishers, 1979; M. Simai and K. Garam(eds), *Economic Integration: Concepts, Theories and Problems*, Budapest, Akadémiai kiadó, 1977; A. Nove, H.-H. Höhman and G. Seidenstecher (eds), *The East European Economies in the 1970s*, London, Butterworth, 1982; S. Fischer-Galati (ed.), *Eastern Europe in the 1980s*, Boulder, Colorado University Press, 1981.

    On the general political situation: O. A. Narkiewicz, *Eastern Europe 1968–1984*, Beckenham, Croom Helm, 1986; T. Rakowska-Harmstone and A. Gyorgy (eds), *Communism in Eastern Europe*, Bloomington, Indiana University Press, 1979; S. Fischer-Galati (ed.), *The Communist Parties of Eastern Europe*, New York, Columbia University Press, 1979; R. L. Tökés (ed.), *Opposition in Eastern Europe*, London, Macmillan, 1979; J. F. Triska and C. Gati (eds), *Blue-collar Workers in Eastern Europe*, London, Allen & Unwin, 1981; J. A. Kuhlman, *The Foreign Policies of Eastern Europe: Domestic and International Determinants*, Leyden, A. W. Sijthoff, 1978.

    On individual countries: W. V. Wallace, *Czechoslovakia*, London, Westview, 1977; B. Kovrig, *Communism in Hungary: From Kun to Kádár*, Stanford, CA, Hoover Institute Press, 1979; M. Shafir, *Romania: Politics, Economics and Society*, London, Frances Pinter, 1985; J. F. Brown,

Bulgaria under Communist Rule, London, Pall Mall, 1970; D. Childs, *The GDR: Moscow's German Ally*, London, Allen & Unwin, 1983.
3. The literature on the Polish crisis is now voluminous. The following are representative: M. D. Simon and R. E. Kanet (eds), *Background to Crisis: Policy and Politics in Gierek's Poland*, Boulder, CO, Westview Press, 1981; A. Bromke, *Poland: The Protracted Crisis*, Oakville, Ont., Mosaic Press, 1984; W. W. Adamski (ed.), 'Crises and Conflicts: The Case of Poland 1980–81', *Sisyphus: Sociological Studies*, vol. 3, Warsaw, 1982; T. G. Ash, *The Polish Revolution: Solidarity 1980–82*, London, Jonathan Cape, 1983; G. Sanford, *Polish Communism in Crisis*, London, Croom Helm, 1983; J. Woodall (ed.), *Policy and Politics in Contemporary Poland: Reform Failure and Crisis*, London, Frances Pinter, 1982.

For an interesting discussion of regime, including Soviet explanations of the Polish crisis, see R. Taras, 'Official Etiologies of Polish Crises: Changing Historiographies and Factional Struggles', *Soviet Studies*, vol. 38, no. 1, 1986, pp. 53–68.

2 The period of tension and drift: after Brezhnev

Intention and reality

By the time Brezhnev addressed the Twenty-sixth Congress of the Communist Party of the Soviet Union in the spring of 1981 he was confident about Comecon's future. His attitude to it had been so much more positive than Stalin's, and his intervention so much more successful than Khrushchev's.

> As we know, the decisive sector of the competition with capitalism is the economy and economic policy. At our last Congress, we, like the other fraternal Parties, set the task of further extending Socialist integration on the basis of long-term special-purpose programmes as a top priority. These programmes are to help us resolve the most acute, vitally important economic problems.
>
> At present, they are being translated into concrete deeds. Integration is gathering momentum. The fruits of specialisation in production are visible in practically all branches of economy, science and technology...
>
> What the Socialist countries have accomplished in economic development and in raising the living standard of people amounts to a whole era...
>
> In the past ten years the economic growth rates of the CMEA countries have been twice those of the developed capitalist countries. The CMEA members continued to be the most dynamically developing group of countries in the world...
>
> Life is setting the task of supplementing co-ordination of our plans with co-ordination of economic policy as a whole.[1]

Much had been achieved, and Comecon was on target.

No doubt, allowing for the occasion, Brezhnev was sincere. But the economic position of neither Comecon as a whole, nor its individual members, was quite as rosy as he painted; and comparatively speaking, it then deteriorated somewhat. To be fair, he did remind his audience of 'many new major problems', and called on 'the leaders of the fraternal countries to discuss them collectively in the near future'. That he could go on a few sentences later to say that 'the pillars of the Socialist state in Poland are in jeopardy' shows that he was perfectly aware where some of these major problems lay—in Eastern Europe.

Following the Twenty-sixth Congress, Eastern Europe experienced a strange interlude in Soviet interest. Brezhnev died about eighteen months later; but he had already lost his Comecon drive. His two immediate successors, Andropov and Chernenko, had little more than a year apiece and therefore had no chance to make an impression on the Soviet Union, let alone on Eastern Europe, though Andropov certainly tried with both. In the spring of 1985 Gorbachev began securing his base. At that stage it was impossible to predict how his policies would evolve. He was comparatively young and conspicuously brisk. His approach was innovative; but his statements of policy tended to the efficient rather than the revolutionary, although they were frequently so short of detail as to allow ample room for manoeuvre at some later stage. During most of this five-year interlude, Tikhonov, the Soviet Prime Minister, carried responsibility for Comecon. But apart from lacking push as yet another septuagenarian in the Moscow hierarchy, he was obviously a candidate for the enforced retirement that came late in 1985. In terms of Brezhnev's prescription, the East European leaders could not for a long time be brought together purposefully to set about devising a common policy.

That is not to say that there was a strong movement away from the Soviet Union or from its interpretation of Comecon. At no point did more than a handful even of Poles hope for some kind of breach with Moscow; and at the opposite end of the spectrum there were hard-liners in most East European establishments who were more loyalist than Andropov or Chernenko or Tikhonov. What can be said is that the five years between Brezhnev's Twenty-sixth Party Congress and Gorbachev's Twenty-seventh were a period of tension and drift in which many of the earlier barriers to Comecon integration were allowed to strengthen—and some new ones were allowed to appear, particularly public dissatisfaction with standards of living and the role of Comecon, and especially of the Soviet Union, in producing this deterioration. That does not mean, however, that developments in Eastern Europe were uniform—or that there was no barrier building within the Soviet Union itself.[2] But towards the end there were strong signs of a fightback in Moscow as Gorbachev developed stratagems for Comecon that were part and parcel of his policies for the Soviet Union.

Poland

The most conspicuous developments in the period 1981–6 were in Poland. Jaruzelski's imposition of martial law in December 1981 and his dismantling of many of the reforms of the previous sixteen months put paid to Solidarity as an official political force. He could release its leaders in two batches in July 1983 and July 1984 without fear of seriously reactivating it. This was not merely because he had the hefty support of the police—or the hidden threat of Soviet intervention. Solidarity itself had been overextended; it had embraced most sections of the population and had espoused a wide variety of demands without being able either to formulate a programme or to negotiate or enforce it. Jaruzelski was also intelligent and patriotic enough to tolerate some of the Solidarity reforms and to initiate others of his own, as well as to try to assuage popular economic discontent. Industrial production has recovered slightly; agriculture has been favoured by good harvests. The cost of living is rising steadily; there are still serious shortages, some of them regional; but the food situation in particular is not as catastrophic as it was becoming in 1980. Jaruzelski's talks with Gorbachev were clearly not easy to begin with, but he can now present a better report from Warsaw than Moscow had a right to expect.

Yet the crisis is not over. Solidarity's paid-up membership may have declined from ten million to one, or to a quarter as the authorities would hold. Wałesa and the other leaders may be totally circumscribed in the actions they can take. But a great part of the adult population is still Solidarity-minded; and there is a remarkably active underground press that publishes everything from newspapers to literature, with an eye on long-term education as well as current information. Many of the five million workers said to be in the new trade unions are, inevitably, former Solidarity members, as are some of those in the reviving Polish United Workers or Communist Party. If the new regime is not opposed *en masse*, it is not widely supported either. Some estimates say that about half to two-thirds of the population is either disillusioned or unimpressed or simply indifferent. And there is talk of the inevitability of another crisis in the way that one has followed another with great regularity since 1956.[3]

None of this produces great enthusiasm for Soviet-style communism. That does not mean to say that there is hostility to the principles of socialism or to good relations with the Soviet Union. It is mainly disillusionment with past practice and with recent developments. There is considerable respect for Jaruzelski himself as an honest and earnest man, and for some of his officers and officials. But Father Popiełusko's murder by four security police in November 1984 seriously weakened that respect, despite their subsequent conviction. It also pushed the great mass of the Polish people closer to the Catholic Church and made of it rather more of an alternative than it had been.

One of the most remarkable phenomena of East European history since the Second World War has been the survival and revival of the Church in Poland. Early persecution apart, its views are so much the antithesis of Marxism that they ought not to have survived the successes of socialism in action. But success has been the other way round. The Church, which is putting up new places of worship, has been recognised by the majority of Poles as caring more for their human welfare and for their nation than the supposedly concerned communists. Archbishop Glemp is as anxious as his predecessors to keep the Church out of temporal politics and has done what he could to moderate the behaviour of both Wałesa and Jaruzelski. But the attenuation of Solidarity and the Popiełusko affair have together propelled the Church into a more open and critical stance; and do what it may, it represents the alternative ideology.

Of course, the Church has no wish to be other than a moral force; it does not want to unseat Jaruzelski—still less to attack Comecon. But by its very existence it weakens the country's ability to be a wholehearted participant in the 'Socialist Community' or an enthusiastic Comecon integrationist. The Church is also one of the major supporters of private agriculture and has been trying—so far without success—to win official approval for a scheme to raise Western capital to modernise and mechanise private farms, a policy hardly in line with planning co-ordination from the centre.

However, not everything can be blamed on the Church. Jaruzelski himself allowed the economic reform begun before and during the Solidarity period to continue in a modified form. Some 80 per cent of Poland's agriculture is already private. Recently, in industry, there has been some movement away from directive towards indicative planning; enterprises have been given more autonomy, and their workers—at least here and there—more say in policy and its consequences, bad as well as good; up to a point the market that already guides agriculture has been given a say in determining the output and prices of industry. This has taken Poland a little way in the direction of Hungary and has ranged it against the more extreme forms of planning. But the experiment has not been entirely successful. Jaruzelski may have been able to collect economists' ideas, and to tell one or two ministers what to do. But resistance from bureaucrats and managers further down the line, very determined not to lose the power that underpins their incomes, seems to have been successful in combining with that of Party colleagues, concerned at the ideological implications of industrial decentralisation, to bring the reforms to a virtual halt.[4] So the great central planners can breathe freely again.

But reform also ran into a series of more nearly economic problems. Many highly skilled managerial and technical staff emigrated during the Solidarity period. Immediately following martial law workers were not willing to pull their weight, which made industrial recovery slow. Groups of them in particularly pro-Solidarity areas such as Nowa Huta or Wrocław were tempted by higher wages, but this has acted as disincentive elsewhere and an inflationary factor at serious odds with the need for additional investment. Import substitution, essential to save foreign currency, necessary to service and repay debts to the West, and to replace Western goods, has reduced both productivity and quality and therefore the competitiveness demanded in hard currency exporting. There has been no money for improvements in transport (not least by road) and storage, which has aggravated shortages of fuel and especially of food. The infrastructure has continued to deteriorate, so that basic services like water and sewage disposal are now at risk. Health conditions have been worsening for years, so that Poland now has an unlikely characteristic for the late twentieth century, though also to be found in the Soviet Union, a rising death rate.[5] A sudden upturn in the birth rate has only added to the demand for social expenditure. The planners may have got back some of their chance of planning; but what they have to contribute to Comecon is not an economy bursting to expand production, let alone to consume less energy and raw materials.

There seems in fact to be a kind of stalemate between the surviving economic reformers and the re-emerging orthodox centralists. Which way it is broken will depend on various external factors, but also on the balance of internal social and political forces. The elderly can find life particularly hard. Yet, for Eastern Europe, Poland remains as it was in the Solidarity period, demographically remarkably young. Those with no memories of the principles of the 1940s and the 1950s are in the majority, that is, those who were brought up on the rising expectations of the 1960s and 1970s and then confronted with shortages, frustrations and betrayals. Some slip abroad, some rescue what they can of broken promises through the second economy where, at a price, they can buy some of the comforts of Western life. Others improve their education, support the Church, and hope for political reform. Some of the younger turn to crime, as some of their elders take to drink. But there is a slimly surviving hope for political reform. The government allowed second candidates to be selected for election to local councils in 1984 and to the *Sejm*, or parliament, in 1985; and though few of them were successful, the voters did not provide the hardliners with the vast majorities they hoped for. In 1982 the government initiated a Patriotic Committee for National Rebirth, intended to provide a forum for the great bulk of the politically middle-of-the-road, but it failed to attract sufficient support to be effective. Yet a new attempt is being made, so far without great success, through it and a new organisation called Consensus to bring Marxists and Catholics together to discuss their views and hopes and to create a middle ground. On the other hand, officials and teachers in higher education are being subjected to political harassments as a means of curbing possible dissent. So optimism is at a premium; and if pessimism triumphs, the young will simply wait for another crisis—which will hardly help the CMEA planners.

Czechoslovakia

It remains to be seen whether Jaruzelski can create a communist ruling group and steer it on to a reforming course that will win it at least a modicum of popular support and enable it to forge an upturn in the Polish economy. He is certainly credited with being a very able strategist who concentrates on essentials. In Czechoslovakia, on the other hand, there is no doubt about the supremacy of the communist elite; but its popularity and its economic success are very much in question. In the last few years, the usefulness of Czechoslovakia's contribution to Comecon has declined.

Dubček and his reforming colleagues were unlucky in 1968 that the Soviet Union was not distracted by a war in Afghanistan—as it was when Solidarity exploded on to the scene in 1980—and that it had Brezhnev at the energetic beginning of his career, not at the end. But the distinguishing features of the Czechoslovak events were that a communist government was putting reforms into practice from above, not simply crumbling before complaints from below, and that far-reaching measures of which the Soviet Union disapproved would in this way become impossible to reverse. Hence the invasion—and its consequences. The Slovaks were bought off with a measure of autonomy under a federal system; the population at large was encouraged to enjoy a higher standard of living; and there were no large-scale show trials. But all the reformers were displaced; the reforms were aborted; dissenters were harassed; and firm rule from the centre was reimposed. Since the reform movement had only been becoming widely popular when it was crushed, since the invasion was so sudden and complete, and since resistance was not part of the Czechoslovak tradition, as it has always been of the Polish, there was no simmering discontent through into the 1970s, despite the outspoken courage of a few Charter 77 protesters.[6] Yet the last few years have witnessed possibly significant changes.

In the first place, a regime dependent upon economic success for its political stability has run into serious problems. Heavy industry is energy-hungry. Domestic coal supplies are running low; Soviet oil is becoming much more expensive. Iranian oil became unobtainable; that now imported from Libya demands hard currency, or its equivalent in sensitive arms exports. A long-term nuclear energy programme has encountered building delays endemic in most of Eastern Europe. Most economies in energy consumption necessitate new equipment. Even where investment moneys are available, some of the equipment has to be imported; and energy-intensive exports are more expensive and therefore less attractive purchases even for the Soviet Union and other Comecon countries, while Czechoslovak trade with the West is declining. Economies also require different working attitudes, and these are equally different to inculcate when consumer costs are rising, not least because of the energy dilemma.

Czechoslovakia deliberately avoided the scale of Western indebtedness that others accumulated in the 1970s and could conceivably get enabling Western loans fairly easily now, except that that would run counter to its generally hardline policy. Official appeals to better disciplines at work, associated with Obzina, the Minister of the Interior who was elevated to a Vice-premiership in 1984, have not had much success in a country where the second economy has become a necessity rather than a luxury and therefore takes precedence in popular priorities. The reinforcement of these appeals following Gorbachev's campaign seems unlikely to have much effect to begin with. The last year of the Five Year Plan was not a good one.

Environmental pollution has also damaged the Czechoslovak economy. Austrian and German complaints about the ravages of sulphur dioxide from factory chimneys in the Czech lands and in Slovakia have now been matched by an awareness of the damage inside the country itself to agriculture and forestry and to the health of people in certain areas. Although, following Hungarian complaints, more resources now seem to have been directed to anti-pollution measures, a deteriorating environment not only imposes an additional burden on the economy, but reduces what little enthusiasm there is for the present government.

This is perhaps the second change affecting Czechoslovakia. For some time Husák and his colleagues seemed to fear the spread of the Polish disease. But the natural caution of the people and the entrenched authority of the state combined to maintain a public attitude of enforced acquiescence. Yet economic malaise has now induced disquiet. A few measures designed to increase the impact of profits and markets on production have had little effect and have not amounted to what the public would consider as reform. Husák is now well into his seventies and past the limit that Gorbachev's accession has set for communist leaders. He obviously cannot last; and many of his colleagues are mentally, if not physically, as old. Despite great efforts, the Czechoslovak authorities have not managed to reduce the age level of the Party as much as they needed to, or to recruit adequately across the occupational spectrum, although maintaining a proportion of membership to total population almost twice that of the Soviet Union.[7]

There is therefore a longing for change; and young people are impatient for it. Not of course that they expect much of the Party, in which they see too much corruption and careerism. Some now look to the Catholic Church which, particularly in Slovakia, has rediscovered its voice. Others seek outlets in adaptations of Western pop culture.[8] Few look to Gorbachev for inspiration. Yet many clearly hope that he will be the stimulus for change, a change that will give Czechoslovakia the opportunity to get new leaders and a course more suited to its own needs. They do not oppose working with the Soviet Union and co-operating with Comecon; but they see sensible internal reforms as a precondition, and one that Gorbachev ought to accept in the interest of having a Czechoslovak economy that therefore works well and makes a positive contribution to the 'Socialist Community'.

Whether their hopes will be fulfilled is another question. But the campaign that the Soviet Union launched before Gorbachev's time against cruise and Pershing missiles in Western Europe had the unintended effect inside Czechoslovakia of rousing opposition to Soviet missiles. Gorbachev seems, from the evidence of his approach to arms talks, to have taken the point. And certainly many Czechoslovaks acquired a new sense of the strength of a cohesive public opinion.

Poland is a difficult country for Gorbachev to handle. Czechoslovakia is not all that much easier.

It may at first sight seem an inhuman thing to say that the Czechoslovaks were unlucky in 1968 in not experiencing the great transformation that Hungary did in 1956. But whatever Kádár's role may have been in the disastrous events of that year, the purge of Stalinists certainly left him free in due course to build a new Hungarian policy. Again, whatever Husák's role in Czechoslovakia in 1968, he was left in subsequent years surrounded by many of the Stalinists whose presence had provoked the Dubček reforms and whose survival in effect stifled them, and still does in spite of some technocratic promotions. Twelve years after Budapest, Kádár introduced his New Economic Mechanism. Eighteen years after Prague, Husák is simply working out his time, even if there are minor signs of change.

Hungary

In many ways what the Czechoslovaks would like to achieve—as would the Poles—is a situation resembling that of the Hungarians. Yet the Hungarians tend to be increasingly dissatisfied. One reason for this is that their real standard of living has been declining now for several years. More and more people, especially in low-status trades or professions, have to find second incomes to meet rising prices; and groups such as pensioners, who depend on allowances from the state, are frequently below the poverty level. Hungary is suffering from 'stagflation', a state of inflation without development; and it hurts. Western debts have to be serviced, and exports are harder to sell. Soviet oil and raw materials cost more; and Hungarian goods are frequently not paid for at a fair price. Complaints are general and frequent, no matter how attractive Hungarian shops appear to other East European visitors.[9]

What others see, too, are far-reaching economic reforms, with production decisions devolved to enterprise level, with prices adjusting to the market, with small-scale private enterprise flourishing in many services, restaurants, shops and building firms, with workers remaining in their factories after hours to produce for their own profit, and with Hungary accepted into the International Monetary Fund. Yet Hungarians see devolution as overly restricted and complain about ministry officials and bank and enterprise managers fighting against change. They point to a curtailed market in which overcapacity in metallurgy or refining is protected against the demand for transfer of scarce resources, for example, to the consumer-durable industries. They resent the low employee limit on private enterprise (at present ten). And at least a significant portion of the professionals among them argue beyond mere membership of the IMF and call for full convertibility.

The political atmosphere in Hungary, however, is better than elsewhere in Eastern Europe. It is not merely that real economic reforms have been introduced, but there is reasonaby open discussion about their fate—whether they have gone too far, or should go further—and not just an intra-Party debate.[10] Trade unions are allowed to bargain on wages, and workers frequently elect factory managers. Dissidents are not numerous and they are occasionally harassed; but they are not relentlessly pursued. In the parliamentary election in May 1985 only 35 seats out of a total of 120 were specifically reserved for special candidates. In other constituencies second candidates were permitted; and, despite some interference, a number were successful, in five cases defeating Central Committee members. And the new parliament has already shown that it has a mind of its own. This represents perhaps the most radical advance in the direction of popular involvement in government anywhere in Comecon.

Yet there remain definite and accepted limits. In the first place, there is no challenge to socialism, simply a discussion on how to make it more effective and equitable. Equally there is no challenge to the leadership of the Party, just a slight experiment in improving its rule. The Party is concerned that it is ageing and attracts more careerists than able people, and it is also aware of the need to siphon discontent into constructive action. Giving non-Party members the chance to make contributions through parliament is one way of improving government without surrendering authority. It also undermines protest. In addition, encouraging the trade-union movement to bargain about wages keeps it out of politics and gives it some responsibility for unpopular economic developments—while in fact it has a mainly negative role in decision-making, and not a particularly effective one. The idea of 'enterprise democracy' in factories is all but dead.[11]

All this, plus the refusal of the last Party Congress in 1985 to agree on a further burst of reform, has tended to strengthen both cynicism and escapism. Those who can earn money—and there are entrepreneurs and artists who earn fabulous sums by the normal standards of Eastern Europe—have fine houses, eat well, dress smartly and drive fast. The majority at least manages to keep up appearances, and hopes for better times. And a minority remains optimistic, takes such part as is possible in politics, and argues in favour of enlarging the function of the market, not just in Hungary, but in Comecon, and in competition with, not isolation from the West. For rather different reasons from the Poles or the Czechoslovaks, the Hungarians are not enthusiastic advocates of Comecon integration. And as Kádár becomes more cautious and less nimble with age, as no obviously enlightened successor emerges from among the contenders, and as even the middle generation in the Party seems conservative, there is a certain lack of confidence in the future.

Yet if some of its architects regard the economic reform in Hungary as less successful than they had hoped, that does not imply that it has been a failure. There is a great deal of prosperity—and of contentment—in the countryside, for Kádár knows how to handle the peasants. If the citizens of Budapest are unhappy, it is partly because they have not yet found a corresponding niche in society to the one their country cousins have occupied for generations. And if the technocrats and intellectuals are unsettled, it is not mainly because of their living standards. To the other East Europeans, Hungary remains exceedingly prosperous—a fact Hungarians themselves frequently use against their neighbours. Their response, for example, to Romania's abuse of their fellow nationals in Transylvania is quietly to enjoy the discomfort that Romanians have suffered during recent hard winters, almost without the energy to light their homes, let alone heat them.

Romania

That does not make the Romanians love them any more, even if still envying them. But Romania is a very special case in Eastern Europe. It has had no insurrection on the Polish, Czechoslovak or Hungarian models, though in the 1950s and 1960s its leaders engaged in fairly dramatic internecine struggles. Perhaps this has been because it is socially and economically less developed. About half its population still lives in the countryside, and it currently styles itself an 'industrial-agrarian developing country'.[12] Or perhaps it has been because it is less fully exposed to Western ideas; contact is indirect and travel is restricted. The Communist Party has also gone to great lengths to prevent contamination from elsewhere in Europe by absorbing the outward forms of change in Czechoslovakia or Poland while strengthening its own monopoly of power. And certainly, for a generation now, Romania has been shepherded by one man, Ceauşescu, who has shown consummate skill in manipulating trends in politics to secure and maintain supremacy over the Party, and who has not only completed the transfer of governmental power to the Party but has appointed members of his extensive family to most leading positions in both the Party and the state. Dissent, let alone disaffection, has become more or less impossible. And unlike the situation in Poland, the Orthodox Church has neither the following nor the tradition to allow it to act in ideological opposition to the state.

In this sense, allowing second and occasionally even third candidates to contest single constituencies in elections to the grand national assembly and the people's councils is a virtual sham compared with elsewhere in Eastern Europe. Even a *coup* allegedly attempted by the army in 1983 had no success in trying to shift Ceauşescu's policies. And encouraging those who farm one-tenth of the land privately is not a reform either. It is a policy expedient, intended to help Romania recover from a disastrous food situation. And it does not use the carrot of profit in the Hungarian fashion, but lays down minimum levels of production over a wide range of products, failure to meet which can result in the loss of private plots. An even more draconian policy is applied in the mines, which have been regimented in face of the miners' discontent and the great coal shortage. The pressure exerted on young people to have bigger families to arrest demographic decline is beyond anything elsewhere.

It might be asked why, in face of what could be described now as a resort to Stalinism or paternalism, there is no serious public unrest. There have been one or two ugly incidents in recent years, despite Ceauşescu's tight control. Part of the answer is artificially organised mass support. But since Romania's output of oil began declining in the 1970s and it became more expensive, too, to import oil (including from the Soviet Union which, in the case of Romania, does not give preferential prices), a drive for energy saving has added impetus to the earlier drive for industrialisation that was the reaction to Khrushchev's plan for a division of labour within CMEA. This has made life extremely hard, but perceptibly in the past four years it has also led to a modest rise in both industrial and agricultural production which at least holds out the prospect for a better life to come.

But another ingredient in the Ceauşescu recipe is his continued exploitation of nationalist sentiment.[13] Keeping Soviet troops off Romanian soil, dallying with the People's Republic of China, castigating the Magyar minority, paying off its indebtedness quickly to spite its Western creditors, and also having its own individual agreement with the Common Market, all these posturings pander to the Romanian sense of being different from their neighbours ethnically and threatened by them politically, above all by the Soviet Union. It is one of the oldest devices in the autocrat's handbook. It is also one that, on balance, the Soviet Union has been prepared to accept. For there is no way in which Romania can defect from the Warsaw pact, still less provide China with a base on the Black Sea (if it wanted one—which it clearly does not). And Ceauşescu presents the Soviet Party with no kind of ideologically subversive threat.

Of course, things could change. Even a moderate reformer like Gorbachev could object to an embarrassing concentration of authority in the hands of one of his socialist colleagues. Ceauşescu's health may not last, and the succession cannot be guaranteed despite the family's strategic position; in the past one or two of them have had to be retired for apparent incompetence. Apart from either of these possibilities, the process of modernisation will produce social change and may provoke protest of a kind that cannot be stifled or gainsaid.[14] Condemning everyone's nuclear arsenal is popular, but also gives ordinary people the idea that they are important. Even within the higher ranks of the Party there are said to be opposition elements. All of this makes Romania potentially unstable, and, in Comecon terms, not the rosiest of prospects. The Soviet Union has enough difficult states to handle. But from the Comecon viewpoint the present alternative is not much better. The Ceauşescu family has perforce to trade more with the Soviet Union now than ten years ago, if it is to import fuel. Yet the key to its public legitimisation is its anti-integration stand. So if it survives, and in order to survive, it must adopt a position of hostility to fresh proposals for co-ordination.

Bulgaria

For its part, Bulgaria gives the appearance of being the East European country least disinclined towards the CMEA. Sitting on the fringe of the 'Socialist Community', it seems nevertheless the most loyal both to the Soviet Union itself and to its special brand of Marxism. It welcomes the Red Army on its soil and hosts top level WTO meetings. Though well into his seventies, Zhivkov, the Party leader, still presides over a system of Party, government, police and army that Brezhnev would recognise and approve if he were here, with no serious personal challenges and little genuine popular participation. He has squeezed out the cultural variety that Lyudmila Zhivkova, his daughter, encouraged in the early 1980s in the last few years of her life, and he has no dissident movement to deal with. The Orthodox Church is as reconciled to the authority of the state as in Russia. And there are enough potential successors around for Zhivkov to be able to keep them all at bay, as he has done in the past, and yet for one of the younger technocrats among them to succeed him in due course.

One source of Zhivkov's strength—and his power has stretched over thirty years, longer than any other East European leader—has been strong support from Moscow in return for equally strong loyalty. But another was his early conversion to consumerism which was then supplemented by bouts of economic reform. The standard of living in Bulgaria is relatively high, and consumer durables are available in a way that they are not in Romania or even in Poland. Food is in good supply and varied. Since 1979 a New Economic Mechanism has been revitalising production. The number of mandatory controls on enterprises has been reduced, and some decisions have been devolved. Steps have also been taken to involve enterprises fairly directly in foreign trade. In the 1970s useful links had been forged between industry and agriculture on a regional basis; and in 1984 their successful incentive and bonus schemes were much improved. And, however unwillingly, the peasants' liking for private plots has been indulged.

But consumerism is a dangerous game. Bulgarian economic growth has tailed off. The economic reforms have not been far-reaching enough to get rid of industrial inefficiency (especially in fields such as metallurgy). Industrial enterprises in general are too large to be cost-effective, and light industry is under-funded. There is a shortage of labour. And the weather, summer and winter, has recently been unkind to agricultural production. Raised expectations are not being met; and corruption has become more widespread as a result. The political reshuffle of 1984 seemed to take some account of this, since a number of promotions to the Politburo were young technocrats. Admonitions to better work discipline have mirrored the Soviet approach but have not so far produced any better results. A new super-ministry is being established to supervise industry and, on the Gorbachev theory, to reduce bureaucracy. But a more fundamental problem is now widely admitted, one that calls in question the relationship with the Soviet Union.

Bulgaria is particularly dependent on Soviet oil. One reward for its good behaviour was to be allowed to resell some oil on the hard currency market. Another was to be able to use oil profligately.

Bulgaria has insufficient coal, and no more than 20 per cent of its electricity is hydro-powered. Now that Soviet prices are being increased, there is no more the possibility of selling abroad, and production costs at home are rising rapidly despite energy-saving initiatives. This has been quickly reflected in the consumer market in higher prices for scarcer petrol and electricity—and for industrial goods and food. The Soviet Union is simultaneously asking for better quality Bulgarian exports in payment for its oil, a demand that undermines Bulgaria's ability to make good its deficiencies by shopping around in the West and in the Third World.

Not every winter will be as hard as that of 1984–5; but the energy crisis is not a matter of a single season. There are, of course, popular distractions that leaders can exploit. A fifth of the population is made up of minorities, of which there are more than fifty, one or two of them with a higher reproduction rate than the ethnic Bulgarians. Anger has in the past been turned against the Turks and Greeks, and there have recently been signs of a revival of this in the case of the Turks. The mixed population of Bulgarian Macedonia, and their connections with Yugoslav Macedonia, give Zhivkov an easy opportunity of scoring nationalist points by sniping at leaders in Belgrade and Skopje. But the Bulgarian population is socially and intellectually much more sophisticated than it was in the good old xenophobic days; and it even has cause to admire some of the independent attitudes that its neighbours can adopt. The main trouble is that, when consumers' complaints come home to roost, they settle on their own regime's shortcomings—and on the Soviet, or the CMEA, connection.

Whereas Romania participated in the 1984 Los Angeles Olympics, despite the Soviet-led boycott, Bulgaria was the first to follow the Moscow line. But that does not mean to say that the Soviet Union can automatically depend on Bulgarian support. In principle, Bulgaria accepts the Soviet view of Comecon. But that view has changed several times, and so now have circumstances. The recent decline in world oil prices could change them again by making Soviet oil cheaper or Middle Eastern oil an attainable alternative. In any case, the Bulgarians like to get on with the business of pursuing their national development by concentrating on improving their own economy rather than on wasteful international issues.[15] So in dealing with Bulgaria, the Soviet Union will have at least a minor problem where previously none existed. Nothing is easy, not even for superpowers.

East Germany

East Germany is a strange paradox. In one sense it is the least East European of the East European states, and therefore the one most likely to be disaffected with Comecon. If, like the Hungarians and the Romanians, the East Germans are not Slavs, they differ in having fellow nationals in a very powerful national state just across their border. Despite forty years of separation since the Second World War, their culture is little different, and their nationalist aspirations, however sublimated in socialist terminology, are in fact much the same.

On the other hand, East Germany depends for its existence on Soviet support, much more than any of its fellow East European states. It is legitimised by the nearly 400,000 Soviet troops on its soil who guarantee its separateness from West Germany and prevent it being sucked into the capitalist attractions of the West. It is thus in no position to argue with Comecon.

Alternatively, it is its leaders who are in no position to argue. While the Soviet Union theoretically exends to East Germany the same recognition as it does, say, to Czechoslovakia or Bulgaria, it retains the rights it acquired by the victory over Hitler's Germany in 1945 and every so often reasserts them under 'four-power authority'. This has not prevented it from forcing the communists in East Berlin to take an entirely different tack. When Moscow responded favourably to West Germany's *Ostpolitik* after 1969 but found Ulbricht unwilling, it simply replaced him in 1971 with his more compliant deputy, Honecker.

In contrast, the East German people have often enough shown their distaste, or at least their dissatisfaction, for the great 'Socialist Community'. Writers and artists have frequently abandoned their homeland. But for the Berlin Wall thrown up in 1961, East Germany might have been emptied of much of its population. As late as 1984 some 40,000 East Germans were allowed to emigrate to West Germany, no doubt only a fraction of those who wished. And a peace movement, which was originally officially encouraged in 1983 to oppose the stationing of cruise and Pershing missiles in West Germany, in due course adopted an officially discouraged opposition to the stationing of Soviet SS-22s in East Germany and, significantly, was joined openly by both the majority Lutheran Church and the minority Catholic.

That is not to say that there is no widespread support for the East

German regime and the Soviet alliance. The Socialist Unity Party, as the Communists are called, probably comprises about 17 per cent of the adult population, and there are many professionals as well as bureaucrats, officers and athletes—and women—who, at worst, have a vested interest in the continuation of the status quo, and who no doubt have more loyalty and enthusiasm than that.

In truth, of course, the paradox is more apparent than real. Since the Basic Treaty between the two Germanies in 1972, East Germany has in fact been an East European country, similar to the others, with Soviet and international recognition as such. But it has continued to have special relationships with the Soviet Union and with West Germany. And whereas, for different reasons, the hardliners at the top and the mass of the population at the bottom originally opposed this strange ambiguity, both groups have since learned not only to live with, but to profit from it. Simultaneously, if there were two groups, they have come together at times, along with those in the middle, to make important but not fundamental criticisms of Comecon and the Soviet connection generally.[16]

The main benefit from the West German link is unashamedly economic. The Federal Republic is anxious to assist the Democratic Republic on humanitarian grounds—fellow nationals suffering across an artificial line—and on political grounds—the two Germanies must some day be one, and in the interim the standing of West Berlin must be protected. In the prevailing international situation its only weapon is economic; per head of population it is more than twice as rich. So it is prepared generally to extend privileges such as loans, and occasionally to pay ransoms, for refugees for example. East Germany therefore finds itself with a more sympathetic partner in Western trade than any other East European state could possibly hope for. It also has more or less open access, at little cost, to West Germany and, basically therefore, to Western technology (though not all the flow is one way). And above all, it has duty-free, and quantitative restriction-free, entry to the Common Market, a privilege which other East European states cry their hearts out for in vain.

This has not led to dependence upon West Germany and so it has remained acceptable to Honecker and the Party elite. It has also helped to make East Germany, per capita, the richest of the East European countries, including Russia. This has delighted Honecker who, as a comparatively successful consumerist, need not feel unduly threatened politically. The populace at large is not unhappy and, in view of the tangible rewards, is prepared to work harder than most East Europeans to increase industrial and agricultural output. So there is the additional bonus that the East German economy is the most productive in the Socialist bloc.

The main benefit from the new form of Soviet link is in standing and respect. The Soviet Union sees East Germany, in consequence of the Basic Treaty, as a channel to West German and Western technology and as a lever for West German and Western opinion, a means of improving considerably its own economic performance and its international capability. So it is only in the last resort that it behaves towards East Germany in 'four-power' terms. It also respects East Germany's economic development and does not object to economic reforms that contribute to this without producing irresistable popular demands for risky political changes. So the East German government is allowed to play a slightly superior role to that, say, of Romania or even of Hungary.

There remain, however, internal differences between much of the Party and some of the people. At the official level there is genuine concern at overexposure to West German ideas, whether through the exchange of visitors or through the medium of television. At the popular level there is frustration among old and young alike at the lack of free movement to and from the West. There is also envy of West German living standards as so clearly portrayed both by visitors and on television, and mounting annoyance that economic reform has gone nothing like so far as in Hungary and resembles much more the dull old model of heightening work discipline currently advocated by Gorbachev.

The fact is that West German economic support facilitated the economic growth of the 1970s and the early 1980s, but that East Germany is now at the point where, in face of decelerating growth and rising prices, it must change from extensive to much more difficult intensive development. Some of this can undoubtedly be achieved through co-operation or integration within Comecon; and East German industrial combines fit Gorbachev's prescription. But for one thing, much greater flexibility is required than is possible with an excess of centralised planning.[17] And for another, Soviet policy on oil has plunged East Germany into difficulty at exactly the wrong time, propelling it towards West Germany for rescue.

Cheap Soviet oil assisted East German economic growth. In due course a rise in price brought home the need for energy-saving and underscored the wisdom of modernising. What hurt was the sudden cut in supplies in 1982 at a time when East Germany was already heavily in convertible debt. Buying fresh hard-currency oil was almost out of the question, at least until West Germany intervened with further loans. Since then East Germany has reduced its imports and expanded its exports to get its trade more in balance. But since then, too, the Soviet Union has stepped up its demand for high quality goods in exchange for oil, precisely the kind that might otherwise earn Deutschmarks or dollars. It is hardly surprising that what was once a common understanding on the value of both the Federal German and the Soviet connections is now somewhat less strongly held in East Germany, and that Honecker, too, seems to have his doubts about Soviet friendship.

This does not make Honecker an anti-Soviet hero. He may at times be popular; but that does not make him a reformer, at least in the radical political sense. In fact it is said that, in fear of what was happening in Poland, he argued in favour of intervention. It was not just the danger of trade unionism spreading, but of East Germany being cut off from Moscow. However, it is unlikely that he was any more enamoured at the siting of Soviet missiles on his territory than many of his countrymen; it was too much like getting back into the pre-1972 front line. And when, towards the end of 1984, he tried to pursue a slightly independent foreign policy line, he was quickly slapped down by the Kremlin. The East German leaders may wish to be loyal and co-operative, and the public has probably no choice. On the other hand, in looking for genuine Comecon integration, Gorbachev may have at least as big a problem with his East German colleagues as with any others.

East European attitudes to Comecon and the Soviet Union

Over a five-year period some resentment inevitably built up towards Comecon for its apparent failings. Standards of living in Eastern Europe mostly declined; and where they did improve marginally, it seemed to be none of Comecon's doing. Indeed, the Comecon link appeared to be damaging. Poland's neighbours, for example, carried some of the cost of helping it to recover from the shortages of 1980–1; and ill-feeling was strong between Czechoslovakia and Hungary over the vexed question of a joint Danube power scheme. Inescapably, much of the resentment was directed against the Soviet Union—as the dominant partner in most co-operative enterprises and, above all, as the supposed friend that was nevertheless causing economic distress by raising energy and raw material prices. The fact that not all of the resentment was justified—consider, for instance, how long Eastern Europe enjoyed the benefit of low-priced oil—made precious little difference. And the Soviet Union is at least as convenient a scapegoat for East Europeans as the United States for West Europeans.

Another economic grievance that built up concerned military expenditure. The Warsaw Treaty Organisation is quite separate from Comecon. It was founded six years later, in 1955; though one or two of its actions have sometimes appeared to overlap with those of Comecon, it has a different purpose and a separate structure. But to some extent its development from a cipher to an organisation that the Soviet High Command wants to capitalise on closely parallels the Comecon story; and it does, of course, make heavy economic demands on its member countries.[18] The greatest burden is on the Soviet Union itself, whose defence expenditure as a percentage of gross national product seems to be twice that of East Germany and five times that of the remaining states. It so happened that the Soviet naval and missile expansion of the 1970s made much greater inroads than expected into total expenditure at a time when Soviet industrial growth was tailing off and agriculture was running into a series of disasters. In 1978, therefore, at a meeting of the WTO Political Consultative Committee, the Soviet government requested that other governments should increase their defence expenditure by 5 per cent. The cost of fighting in Afghanistan from 1979 on, and of conducting a missile race with the United States, added insistence to the request. At the same time, in response to a NATO programme, a start was made on modernising conventional equipment in the East European armies, which raised the prospect of yet higher expenditure. How great a percentage rise the East Europeans actually conceded is not clear. But the Romanians walked out of the meeting and refused to pay more; and the remainder at least resented what they were asked to do and may not have paid very much extra. The breakdown in East–West arms talks in 1983 and the prospect of a military race in space were viewed with horror; and the resumption of the talks in 1985 was greeted with a relief which still did not remove the basic concern about the burden of Soviet defence demands.

Economic dissatisfaction, including that laid at the door of the Soviet Union, is now rather widespread. Much of it comes down to fairly obvious basic questions. Certainly there is economic inequality in the United States and social distress in the Common Market. But why, seventy years after the Revolution, is the Soviet Union so low in the industrial and technological league tables? Admittedly there were generations of backwardness to overcome and a devastating war to be fought and recovered from. Yet, why is the Soviet Union so far behind Japan, which had similar obstacles, and why is it possibly being overtaken by China, with no less problems? These are the issues which Gorbachev himself is confronted with, and to which his new policies must be directed. No simple answers come from Eastern Europe. In the first place, there are none. In the second place, few East Europeans have the position—or could have the courage—to tell Moscow what to do to get out of an implied mess. And in the third place, there is disagreement among the East Europeans themselves.

Clearly, for instance, there is no one Polish recipe for economic success; and the New Economic Mechanisms in Hungary and Bulgaria are quite different from each other. Again, what might appeal to the publics in, for example, Czechoslovakia or Romania would hardly coincide with what their governments might recommend, since the satisfaction of consumers' wants would not automatically maintain the elites' monopoly of power. On the other hand, in so far as views are made known, there are two common threads, either that there should be improved central planning, or that the market should play a very much greater role in decision-making processes. In Eastern Europe, however, in patent contrast with the Soviet Union, the improved planning argument tends to stop at national frontiers and merely pays lip-service to Comecon integration; and within each country it concedes elements of the market argument, at least at the peasant fringes. In addition, in the period between the two Soviet Congresses, the main supporters of fundamental change in Comecon economic policies were able to develop their case against a background of East European dissatisfaction and Soviet distraction.[19]

These were mostly Hungarian economists, already in the van of change through helping to initiate the New Economic Mechanism back in 1968 and trying to develop it thereafter. Within Hungary itself, of course, they ran into difficulties in the early years as Kádár resisted change in face of uncertainties about the succession in the Soviet Union—and in Hungary. Yet Budapest decided to join the World Bank and the International Monetary Fund in 1982, and in 1985 to join the International Finance Corporation and the International Development Association. The Hungarian economists went on to argue that, although CMEA had suffered from trading with the West in the 1970s due to world recession, it was essential for it to develop this trade, as well as regional co-operation, as the only way to improve its overall economic performance in the long run. This carried the further implication that, as in Western economies, more regard would have to be had to market forces. There was no point in trying unsuccessfully to buy advanced machinery with overpriced foodstuffs or the inferior products of uncompetitive metallurgical, petrochemical or consumer-durable industries. Economic reform would have to develop further, to disestablish monopolies and to assist the movement of investment from unprofitable enterprises to those that would produce what customers abroad would actually buy—and domestic consumers would incidentally benefit from. This was an argument with unwelcome implications for some politicians, bureaucrats and managers in many corners of Eastern Europe. But at least part of it appealed to national elites anxious to engage in profitable hard-currency trade to dispose of their Western debts and finally to gain access to the technology of the future.

Oddly enough, the idea that Eastern Europe should trade more with the West and the Third World, and thereby become more subject to market forces, was one that had been inadvertently advanced by the Soviet Union when it began to raise the price of oil to world levels and to reduce the supply. Recourse to the world market was also seized upon by quite a number of East European governments as a means of putting their own interests first and resisting further calls to greater integration. Hungarian anti-centralising ideas might have fared worse but for this unlikely assistance. They did not fare well in that they nowhere recruited official support and, in the middle of the period, ran into increasing Soviet opposition as ideologically unacceptable and politically dangerous. But they remain as a challenge to the Soviet system and its view of Comecon's future. Gorbachev faces some awkward situations in Eastern Europe and a certain amount of economic discontent; he also faces an alternative interpretation of existing ideology that might just be more successful economically.

Disgruntlement in Eastern Europe is political as well as economic. Internally it is cleverly contained, as in Jaruzelski's Poland, or is diverted against external targets, as in Ceauşescu's Romania. In no country, and at no level, is it, or can it be, overtly anti-Soviet. Yet in the five years from 1981 to the end of 1985 there were signs of official resentment, too, at Soviet policies. In communiqués following specific CMEA and WTO meetings the East European states indicated their firm support for Moscow's stand on nuclear arms and missiles. But in general their relations with the West did not deteriorate as much as those of the Soviet Union did. There were opposing extremes, Czechoslovakia being particularly anti-American, and Romania condemning both superpowers' intransigence. But Kádár, for example, opened an active link with Britain in 1983 and with Italy and France the following year; and, throughout the period, Honecker cultivated the special relationship with West Germany, easing some travel restrictions and negotiating further loans, and received the Italian, Swedish and Greek prime ministers. East European appeals for the non-deployment of the new American missiles in Western Europe were more than automatic mouthings of Soviet objectives. They represented concern that an apprehensive Moscow might as easily instigate a war as an aggressive Washington. At WTO meetings, too, East European ministers urged the Russians to put the same emphasis on conventional arms reduction talks as on nuclear. Although in 1984 the Czechoslovak government may have welcomed the arrival of SS-22s on its territory, the East German government had no choice, and the Hungarian was relieved that the question did not arise in its case; but there were public protests in both Czechoslovakia and East Germany. Very soon, too, Bulgaria joined Romania in proposing a nuclear-free zone in the Balkans. The WTO was supposed to renew and revise its treaty at a meeting in Sofia in January 1985; but the function was delayed till April and led to a simple renewal, without elaboration.

None of this was really a threat to Moscow; and some of it resulted from mistakes and confusion there. Yet there were unusual aspects to East European behaviour. Romania had maintained a friendly connection with China since the beginning of the Soviet–Chinese split; so it was not out of the ordinary for Ceauşescu to receive important Chinese visitors. The Poles began to stride ahead of the Russians in developing a new relationship with Beijing.[20] But the real shock was for Sindermann, President of the East German Assembly, to visit China in 1985. For Jaruzelski to continue discussions with Archbishop Glemp, and the Poles in general to look to their co-religionists in the West, was natural in the circumstances. It was a little more radical for Szürös, who was appointed head of the Foreign Bureau of the Hungarian Communist Party's Central Committee in 1983, to begin making speeches about the different shades of socialism and the importance of small and medium states. However, for Honecker to prepare to visit West Germany was carrying the special relationship too far, particularly since he first visited the out-of-favour Ceauşescu and agreed with him publicly on the importance of *détente*.[21] And for Zhivkov of all people, from Bulgaria of all countries, to take steps to emulate him was sufficient to make the Soviet Union in the autumn of 1984 forbid them both. However, Ceauşescu visited West Germany and he continued to develop his relationship with Yugoslavia and Israel in defiance of known Soviet attitudes. Yet there was no outright challenge to the Soviet position. Even Romania was more bark than bite. Nevertheless, apart from

Czechoslovakia, which fiercely lambasted both Hungary and East Germany in 1984, all the East European states were officially demonstrating a marked disinclination for the Soviet Union and its policies, with Romania out in front criticising the Brezhnev doctrine. Alternatively, they were becoming much more interested in Western Europe, and not least in the Common Market, with Romania again to the fore but very closely followed by most of the others. This remains the challenge still, economic dissatisfaction coupled with political disaffection, and a general feeling that the Soviet Union is far too mechanistic in its approach to the complexities and the feelings of Eastern Europe as a whole.

Notes

1. *Documents and Resolutions of the Twenty-sixth Congress of the Communist Party of the Soviet Union*, Moscow, Novosti, 1981, pp. 8–40.
2. Any discussion of East European fortunes in the period 1981–6 must inevitably draw on a wide range of source materials, only a few of which can be specifically referred to in a book of this size. A useful series of specialist articles was published under the title 'East Europe, 1985' in *Current History*, vol. 84, no. 505, 1985, pp. 353–93. Two thoughtful papers on 'Recent Developments in Eastern Europe' were presented to the NASEES Conference in 1984, one by J. Dempsey on 'Political Development' and the other by W. Brus on 'Economic Developments'; similarly in 1986, one by J. Batt on 'Recent Political Developments' and the other by A. H. Smith on 'Recent Economic Developments'; and also in 1986, M. Nielsen on 'Problems Concerning Reforms in CMEA Cooperation', and M. Lavigne on 'Developments in CMEA'. Two useful collections of academic articles are: M. J. Sodaro and S. L. Wolchik (eds), *Foreign and Domestic Policy in Eastern Europe in the 1980s*, London, Macmillan, 1983; and M. McCauley and S. Carter (eds), *Leadership and Succession in the Soviet Union, Eastern Europe and China*, London, Macmillan, 1986.
3. Recent opinion polls are still being analysed. But cf. D. S. Mason, *Public Opinion and Political Change in Poland 1981–1982*, London, Cambridge University Press, 1985. For an interesting discussion of 'economic–political cycles', see J. Staniszkis, *Poland's Self-limiting Revolution*, Princeton, Princeton University Press, 1984, pp. 248–77.
4. For some discussion of Jaruzelski and the Polish economy see Z. M. Fallenbuchl, 'The Polish Economy Under Martial Law', *Soviet Studies*, vol. 36, no. 4, 1984, pp. 513–27; G. Blazyca, 'The Polish Economy Under Martial Law—A Dissenting View', *Soviet Studies*, vol. 37, no. 3, 1985, pp. 428–36; and S. Gomulka and J. Rostowski, 'The Reformed Polish Economic System 1982–83', *Soviet Studies*, vol. 36, no. 3, 1984, pp. 386–405; see also *Polish Perspectives*, vol. 28, no. 2, 1985, pp. 40–2, concerning the position of directors.
5. M. Okólski and B. Puƚaska, *Recent Mortality Patterns and Trends in*

Poland, Warsaw, University of Warsaw, 1982; and M. Okólski, 'Demographic Transition in Poland: The Present Phase', *Oeconmica Polona*, 1983, no. 2, pp. 185-232.
6. Two accounts of this period are V. V. Kusin, *From Dubček to Charter 77*, Edinburgh, Q Press, 1978; and H. G. Skilling, *Charter 77 and Human Rights in Czechoslovakia*, London, Allen & Unwin, 1981.
7. Cf. G. Wightman, 'Membership of the Communist Party of Czechoslovakia in the 1970s: Continuing Divergence from the Soviet Model', *Soviet Studies*, vol. 35, no. 2, 1983, pp. 208-22.
8. H. G. Skilling, 'Independent Currents in Czechoslovakia', *Problems of Communism*, vol. 34, January-February 1985, pp. 32-49.
9. R. L. Tökés, 'Hungarian Reform Imperatives', *Problems of Communism*, vol. 33, September-October 1984, pp. 1-23.
10. Cf. F. Havasi, 'Economic Growth and Equilibrium', *New Hungarian Quarterly*, vol. 26, no. 98, 1985, pp. 18-26, and R. Nyers and M. Tardos, 'The Necessity for Consolidation of the Economy and the Possibility of Development in Hungary', *Acta Oeconomica*, vol. 32, nos 1-2, 1984, pp. 1-19, and B. Csikós-Nagy, 'A New Phase in the Hungarian Economic Reform: 1984-87', *New Hungarian Quarterly*, vol. 25, no. 96, 1984, pp. 21-32.
11. Cf. B. A. Racz, 'Recent Developments in Hungarian Enterprise Democracy', *Soviet Studies*, vol. 36, no. 4, 1984, pp. 544-59.
12. Agerpress, *The Socialist Republic of Romania*, Bucharest, Agerpress, 1985, p. 28.
13. Cf. D. Mazilu, *National Independence: Romanian Thinking and Action*, Bucharest, Military Publishing House, 1984.
14. Cf. D. N. Nelson, *Demographic Centralism in Romania: A Study of Local Communist Politics*, New York, Columbia University Press, 1980, pp. 145-9.
15. Cf. J. R. Lampe, *The Bulgarian Economy in the Twentieth Century*, Beckenham, Croom Helm, 1986, p. 230.
16. Contributions on East Germany are to be found in D. Childs (ed.), *Honecker's Germany*, London, Allen & Unwin, 1985.
17. Cf. P. Boot, 'Continuity and Change in the Planning System of the German Democratic Republic', *Soviet Studies*, vol. 35, no. 3, 1983, pp. 331-42.
18. Cf. K. Dawisha and P. Hanson (eds), *Soviet-East European Dilemmas*, London, Heinemann, 1981, pp. 134-71; on WTO questions generally, see D. Holloway and M. O. Sharp (eds), *The Warsaw Pact: Alliance in Transition?*, London, Macmillan, 1984; and A. Braun, *Small-state Security in the Balkans*, London, Macmillan, 1983.
19. From an extensive literature, see L. Csaba, 'Planning and Finances in the Decade after the Adoption of the Comprehensive Programme in the CMEA', *Acta Oeconomica*, vol. 27, nos 3-4, 1981, pp. 351-72; and 'The Role of CMEA in the World Economy in the 1980s', *ACES Bulletin*, vol. 26, nos 2-3, 1984, pp. 1-27; I. Dobozi and H. Matejka (eds), 'East-West Relations in the Mid-Eighties: In Search of a New Equilibrium', *Trends in World Economy*, no. 47, Budapest, 1984; A. Köves, *The CMEA Countries in the World Economy: Turning Inwards or Turning Outwards*, Budapest, Akadémiai kiadó, 1985.

20. Andrzej Halimarski, 'Relations with China', *Polish Perspectives*, vol. 28, no. 1, 1985, pp. 22–7.
21. A. J. McAdams, *East Germany and Detente*, London, Cambridge University Press, 1985, pp. 193–200.

3 From drift to determination: towards Gorbachev

Soviet priorities in Eastern Europe

When Brezhnev addressed the Twenty-sixth Party Congress he was patently fully aware of political dissatisfaction, at least in certain areas of Eastern Europe. The man whose name was attached to the Brezhnev Doctrine was not likely to be insensitive to developments in Poland and their threat to the *status quo* throughout Eastern Europe and in the Soviet Union itself. So he gave a very specific warning:

> Communists have always faced courageously the attacks of the adversary, and have invariably won. That is how it was and how it will be. And let no-one doubt our common determination to secure our interests and to defend the Socialist gains of the peoples.

At the same time, he was anxious to be accommodating:

> Some time ago the leaderships of a few Communists Parties began vigorously to defend the right to specifically national ways and forms of struggle for Socialism and of building Socialism. But if you look at this without prejudice, you will see that no-one is imposing any stereotypes or patterns that ignore the distinctions of any country ... Our Party has never departed from Lenin's principle which has by now been thoroughly corroborated by the facts of history. In none of the now existing Socialist countries have the forms, methods, and ways of the Socialist revolution been a mechanical repetition of outside experience ... Unless one ignores the actual facts, one cannot speak of any uniformity, or contrast Communist Parties according to the criterion of recognising or not recognising the ways they choose to reconstruct society.

And the Soviet Union was also willing to entertain criticisms:

> Far be it from us to think that everything we have is ideal. In the USSR Socialism was built in incredibly difficult conditions. The Party hewed its way through virgin land. And nobody knows better

than we do what difficulties and shortcomings occurred along the way, and which of them have still to be overcome. We pay close heed to comradely, constructive criticism.

On the other hand, Brezhnev stated very clear limits:

> We are categorically opposed to criticism which distorts Socialist reality and, wittingly or unwittingly, thereby does a good turn to imperialist propaganda, to our class opponent. As our Party sees it, differences of opinion between Communists can be overcome—unless, of course, they are fundamental differences between revolutionaries and reformists, between creative Marxism and dogmatic sectarianism or ultra-left adventurism. In that case, there can be no compromise—today just as in Lenin's lifetime.[1]

In short, the socialism practised by his Comecon partners need not be identical to that in the Soviet Union, but it must be fundamentally the same.

All of this indicated Brezhnev's priorities. He was addressing his East European comrades not as Comecon but as communist partners. His approach to Comecon was far more positive than Stalin's and far more effective than Khrushchev's. Yet it was still essentially political. His drive towards co-ordination leading to integration had followed the invasion of Czechoslovakia and was intended to build a tighter economic framework that would obviate further rebellion and so make counter-action unnecessary. The important thing was to get the politics right; and that meant ensuring that nothing was done that would upset the political balance in the USSR.

The Soviet Union could not allow even one East European people to endanger the supremacy of its own Communist Party since that would put at risk its country's adherence to the 'Socialist Community' and possibly threaten the position or the existence of the Soviet Communist Party. To challenge the leading role of the Communist Party anywhere in Eastern Europe, by tolerating other political parties, or allowing trade unions or other special interest groups to assume political functions, would be to undermine the support of the local elite and threaten the survival of the Soviet elite. A similar misfortune might result from an attack on democratic centralism, or on central planning, if it led to Party or government decisions emerging from below, not as at present being inspired from the top.

There were at least three further worries. The Soviet Union has more than a hundred nationalities. Barely half its people are Russian; and the nationalities living to the east of the Urals are growing much faster in numbers than those to the west. It had therefore become a matter of serious concern for the existing Soviet elite to suppress nationalist ideas, which are in any case contrary to the beliefs and injunctions of Marxism–Leninism. This made it imperative to prevent the division of the Czechoslovak Communist Party in 1968 into separate Czech and Slovak parties, and made it necessary in the early 1980s to seek a solution of the Polish question that could be seen to be communist, not nationalist. Secondly, the Soviet Union has a sizeable Muslim population (which, rather uncomfortably, is also non-Russian). So in seeking a political settlement in Poland, it could hardly concede one that seemed to yield a political role to the Catholic Church. And finally, economic devolution of the Hungarian type, translated to the federal structure of the Soviet Union, could lead to awkward demands for political devolution. Moscow's approach even to mild schemes for economic decentralisation was accordingly bound to be sceptical. When Brezhnev was speaking in the spring of 1981, Czechoslovakia was still fresh in his memory, Poland was still in crisis, and the Hungarian experiment was a continuing talking-point. So it was inevitable that he spelt out a hard political line to the whole of Eastern Europe.

At least equally important in Brezhnev's mind was the military lesson that had to be expounded to Eastern Europe:

> The defensive political and military alliance of the Socialist countries is faithfully serving the cause of peace. It has all the requisites reliably to defend the Socialist gains of our peoples. And we will do everything for this to be so in the future.

In one sense the Soviet Union had never been stronger. By 1981 its nuclear throw-weight had overtaken that of the United States. It at last possessed a powerful navy with overseas bases, so that it could never again be humiliated by the United States as it had been at the time of the Cuban Missile Crisis in 1962. Yet it had to fight a short but uncomfortable border war with China in 1969 and it was now engaged in a damaging struggle in Afghanistan. With West European consent the United States was preparing to instal intermediate-range missiles within easy striking distance of Moscow. And on top of everything else, the loyalty or interest of the Polish army was now in doubt, and the security of land-based communication with East Germany somewhat at risk. This was a moment of crisis in which long-term economic planning was important, but not the top priority.[2]

With the passing of time, too, Eastern Europe had in fact assumed a new importance in military terms. The re-emergence of the intermediate-range missile and NATO's efforts to improve its conventional capability made possession of a defensive–offensive base crucial to the Red Army's tactics in the West. Avoidance of a war on two fronts had been fundamental to Soviet strategy since before the Second World War. But throughout the 1970s a *rapprochement* between the United States and China had developed to the point of diplomatic recognition in 1978 and, since then, to discussions, not simply about trade, but about military support. China had also broadened its contacts with Western Europe, not to speak of Yugoslavia, still less of Romania. This was all the more reason for the Red Army to be concerned to have a secure base in Eastern Europe. In addition, the southern tier of the WTO had become more pivotal in view of the Soviet Union's deeper involvement in the whole of the Middle East and its maintenance of a considerable naval force in the Eastern Mediterranean. In sum, as Brezhnev had raised the standing of the Soviet Union to that of a genuine superpower, he had also heightened the importance of Eastern Europe.

From his point of view Eastern Europe had another, external political importance. 'No-one should have any doubts', he said, 'that the Communist Party of the Soviet Union will consistently continue the policy of promoting co-operation between the USSR and the newly-free countries, and consolidating the alliance of world Socialism and the national liberation movement.' In one form or another, according to the prevailing circumstances in the world, this had been Soviet doctrine for decades. And in the past two decades Eastern Europe had played a key role. Part of this was through the presence of East European technical advisers in Africa and the Middle East, advertising the success of socialism as well as supplying actual aid. But it was just as much what they advertised, in this and other ways, of their relationship with the Soviet Union.

Yugoslavia's independent, neutralist and relatively successful socialism had long acted as a stricture both on the Soviet version and on Soviet attitudes to newly-free countries. Hungary's experience in the 1950s and Czechoslovakia's in the 1960s had reinforced the new states' suspicion of the appropriateness of the Soviet economic model to Afro-Asian conditions and of the wisdom of too great a reliance on Soviet support. Now, in 1981, the Soviet Union had been overwhelmingly condemned by the United Nations for its socialist intervention in Afghanistan. Although Polish engineers and scientists were still doing a highly-skilled job abroad, what the majority could say about Gierek's failure and the subsequent Soviet hostily to Solidarity would hardly be good copy for advertising. In addition, China had already embarked on far-reaching reforms, relevant to populous rural societies, that might well be more successful and therefore more attractive to the Third World and that carried none of the hallmarks of superpower arrogance. Somehow or other, the relationship with the East Europeans, as well as the performance of all their economies, had to be improved so as to impress the poor peoples of the world, without, however, risking the Soviet Union's pre-eminence.

The Soviet Union and the politics and economics of Eastern Europe

In at least one respect the five years that followed the Twenty-sixth Congress proved less troublesome politically than Brezhnev seems to have feared. It is said that intervention in Poland on the model of Czechoslovakia was seriously considered. Marshal Kulikov, Commander-in-Chief of the Warsaw Treaty forces, was a very busy man. There were troop movements on the frontier not far from Gdansk, Wałesa's home territory. Soviet soldiers in Polish uniforms may have played a part in the imposition of martial law; alternatively, Soviet-trained officers may have been considered perfectly adequate. There was certainly immense political pressure, which led to the replacement of a civilian General Secretary of the Party by a military man. But in the event, Jaruzelski forestalled Soviet action by himself assuming dictatorial powers and disbanding Solidarity. This was the kind of self-imposed discipline that Khrushchev had accepted in Gomulka's time and that was not entirely displeasing to Brezhnev. There was no messy fighting as there might have been with Polish patriotism involved; and at least some of the international stigma descended on Polish shoulders. Above all, the Solidarity infection did not spread elsewhere in Eastern Europe.

On the other hand, Brezhnev did not live to see the reverse side of the Polish coin. Martial law was partly lifted by Jaruzelski at the end of 1982 before the Communist Party was reconstructed and while some of the Solidarity reforms were at least being debated. In the spring of 1983 two leading Soviet journals found cause to attack reformers within the Polish Communist Party. But Jaruzelski proved as nimble-footed in dealing with Moscow as with opposition at home, and in the summer of 1983 was able to allow the Pope to visit Poland, much to

Moscow's chagrin. In due course, too, he secured permission to dismantle martial law completely while still presiding over a reform discussion. In the circumstances, first of Andropov's early death and then of Chernenko's ill health, and the resulting internal divisions, it was difficult for the Soviet Union to take effective restraining action, and it had to watch the open trial and conviction of secret police for murdering a priest. However, with the assumption by Gorbachev of the reins of power in 1985, the limits of manoeuvre for Poland were apparently much more clearly defined, so that by the end of the year the process of economic reform, with its political consequences, was almost moribund. Yet Jaruzelski was still in power, perhaps as the best available bet. The Soviet Union might not like all it saw in Poland, but there was as yet little it could do about it.

At the same time, there were other comforts the Soviet Union could draw from Jaruzelski's stewardship by the beginning of 1986. Politically he might seem a little too reformist; but by this very fact he had managed to assuage contrasting critics in the West and in the Third World. Yet he had continued to support Soviet foreign policies in exchanges with Western spokesmen. In part he had no alternative, since the anniversaries of Yalta and of the end of the Second World War renewed American and West German questioning of the Polish territorial settlement. But the reason was less important than the result.

Throughout the inter-congress period Czechoslovakia occasioned no concern. Towards the end there were signs that Husák at 73 might be thought too old; after all, in the spring of 1985, the favoured age for leaders dropped twenty years. Or, if not too old, politically too moribund, and unable to re-energise the Czechoslovak economy. Yet in foreign policy matters, Czechoslovakia was unimpeachable in its loyalty, and a useful critic of the disturbing behaviour of Hungary and East Germany. In 1984, as already indicated, Czechoslovakia made several sharp attacks on Kádár and Honecker for their growing friendliness towards the West at a time when the Soviet Union was locked in diplomatic conflict on the issue of cruise and Pershing. This prepared the ground for the veto on Honecker (and Zhivkov) going to West Germany in the autumn.

In contrast, Hungary did occasion concern. Andropov apparently approved the Hungarian economic model in 1983. But Kádár pursued his innovative foreign policy despite his next-door neighbour's protests, and even entertained Kohl in June 1984 to sound him out about EEC matters. He also paid visits to Paris in 1984 and London in 1985. For his part, Honecker was relieved of interference from the meddlesome Soviet ambassador, Abrasimov, in June 1983. Nevertheless he had to be reminded in May 1985 by Gorbachev himself that the German question was not open, and by Gromyko five months later that there was no community of responsibility between the two Germanies and that damage limitation was therefore not an option. East Germany had to be told to toe the Soviet line, that, in the political imagery of the time, fire would not mix with ice.

In the Bulgarian case the disagreement was the first of its kind, and it preceded Gorbachev's accession—though it followed a trip he made to Sofia on his way to the top. On the other hand, disagreement with Ceauşescu was frequent and difficult to reconcile. Gromyko visited him in Bucarest early in 1984, and Chernenko received him in Moscow in the summer. But he persisted in condemning Soviet intermediate-range missiles together with American, in calling for an end to both NATO and WTO and not taking part in WTO exercises, in proposing a nuclear-free zone in the Balkans, and in receiving American and Chinese visitors. Eventually, even his criticism of SDI was fairly moderate. The most that Ceauşescu would do to please the Soviet Union, but that suited himself, was to take a strong line with the United States on the vexed question of debt repayment. At least, however, he kept his citizenry in line.

In general, the position was such that in June 1985 *Pravda* felt it necessary to print two bitter articles warning all the East European states to eschew a whole series of nationalist, revisionist, anti-socialist attitudes in their domestic policies, and not to try to pursue independent foreign policies towards the West and especially not to try to mediate: Soviet and socialist foreign policies were identical.[3] Even if the strength of this warning may have been connected with the internal struggle in the Soviet Union as Gorbachev secured his position (the journal *Kommunist* soon published a rather different view), it was still an indication of a rather unsatisfactory state of affairs from Moscow's point of view. That the articles may also have been linked with the process of moving towards an East–West summit is additional evidence of how concerned the Soviet Union was about the recent behaviour of some of its East European allies.

By October, however, Gorbachev, now firmly in command, had improved the Soviet position considerably.[4] It was in that month that the Political Consultative Committee of the WTO held an important meeting in Sofia just before the Geneva Summit. In April it had been none too easy to get the Warsaw states to agree to a renewal of the Treaty for another twenty-plus-ten years; and certainly none of the apparent Soviet strengthenings of its text had been accepted. Now, however, Gorbachev secured unanimous support for the stand he proposed to take with Reagan. Given the widespread hope for peace, that was not surprising, but it was still reassuring. Some of the unanimity stemmed from reminding the East Europeans about German revisionism. But Gorbachev clearly conceded a number of crucial points. There was agreement on the need for nuclear-free zones in various parts of Europe and for preventing the development and production of more destructive conventional arms. There was also a declaration of readiness to co-operate with China. This was not a particularly high price to pay, and all of these objectives could with a little care be worked into the new policy that Gorbachev was producing. But the point was that Gorbachev was prepared to exercise that care. Reviewing the meeting later, the Politburo drew attention to Western attempts to differentiate its policy towards the socialist countries, to pit some against the others, cause friction, and hamper their co-operation on the international scene. WTO members, it said, had exchanged views on topical issues of co-operation among them; and 'the prime result' of the meeting was 'the further strengthening of the unity and cohesion of the allied Socialist states'.[5] It was not agreement on everything in line with Soviet views, certainly not on domestic politics, but it was at least a partial recovery. Needless to say, the outcome of the Geneva Summit in November was an enormous relief to Eastern Europe.

The same articles in June criticised those East European governments which favoured economic reforms that weakened centralised planning, propagated market competition and increased the extent of the private sector. No doubt this, too, had a whiff of the power struggle about it. But it followed quicky on the heels of a Comecon session in Warsaw, attended by Tikhonov who, though soon to be unseated by Gorbachev, still represented what was the long-held majority view of the political dangers of economic reforms as practised or propounded in Hungary and elsewhere. The five years between the decline of Brezhnev and the rise of Gorbachev had seen no measurable fall in the interest of Eastern Europe in policies that could destabilise it, undermine the WTO alliance, and even threaten to destabilise the Soviet Union. Even allowing for partisan exaggeration, Gorbachev had no easy inheritance.

The attack on reform was directed against Comecon members and may not have made them any more enamoured of the virtues of the organisation. But it was not necessarily a criticism of their attitude to integration—though it may have been that as well. Brezhnev had proposed the summit to discuss their problems as long ago as the 1981 Congress. But bickering persisted right through to June 1984 when Chernenko and all the other leaders, with the single exception of Castro, sat down to talk in Moscow. There were what Tass called constructive and businesslike discussions and Zamyatin, the official Soviet spokesman, characterised as frank exchanges of views. Certainly there were enough issues to differ about; the levels and prices of Soviet energy supplies; the quality and availability of East European industrial products; and debts to the West. There appears to have been an open split on the question of integration. The Soviet Union itself, Czechoslovakia and Poland favoured it; Hungary, East Germany and Romania resisted it. In the upshot the talk was mostly of co-operation, at best of co-ordination.

At the Havana Council the following October it was specifically agreed to co-ordinate the 1986–90 five-year plans, as those for 1981–5 had been co-ordinated. This was half a victory for Moscow—or possibly a quarter since the whole process was so cumbrous. It was also decided to start work on a comprehensive programme for science and technology to run for 15–20 years. And between then and the next Council in Warsaw in June 1985 a series of long-term bilateral agreements was ostentatiously concluded between the Soviet Union and each of its Comecon partners. Consequently, the major theme of the Warsaw Council was co-operation to overcome the West's lead in high technology, an aim which, if attainable, was bound to be attractive to all member countries. Yet the programme remained less than the Soviet Union might have hoped since it was based on no more than multilateral co-operation. So while the Soviet press congratulated Comecon on its forward-looking achievements in general, it was open to *Pravda* to complain about some of its less helpful members.

One thing that emerged in 1984 and was confirmed in 1985 was that Comecon would not move towards some kind of autarchy, as Tikhonov and some of his specialist advisers had long been advocating, but would continue to trade with the capitalist West, at least where it was to Comecon's advantage. If this was a concession to the East Europeans (and to be frank it was a concession to practicality), it did not go as far as the convertibility sought by the Hungarians. And Soviet spokesmen made quite clear both that intra-Comecon trade should grow and that more of the better quality goods from Eastern Europe should find their way to Russia rather than to the West.

In one respect, the Warsaw Council went further than some of the East Europeans expected. It invited the Common Market to agree to discussions with Comecon designed to lead to mutual recognition. This was the second attempt; the first had been turned down in 1980. On one interpretation, the intention was to increase East–West trade so that the Soviet Union might acquire more high technology and the others more hard currency. Given that most of the East Europeans had been trying to breach EEC trade barriers for years (the Romanians had won their agreement in 1980, and East Germany enjoys a back-door advantage), and that some of them were anxious to avoid increasing their exports to the Soviet market, this looked like a further concession. But a different interpretation was that Moscow was anxious to control all Comecon trade with the Common Market as a means of reducing it and of obviating its possible ecnomic and political consequences. The official approach to the EEC also mentioned the importance of mutual recognition for the process of *détente*—an outcome which, if achieved, would be some compensation to the East Europeans for the bottling up of their trade. So even on the less happy of the two interpretations they did not lose everything. Nor could Russia hope to gain everything.

The initial EEC response was off-hand and negative, but it was being considered again by the beginning of 1986. However, Soviet satisfaction with the CMEA meeting in Warsaw grew stronger as 1985 wore on. The establishment of the new Committee on Co-operation in Engineering took pride of place as the drive developed to bring all possible resources together to catch up on the West's lead in advanced industrial technology. Impressive lists were published of new agreements reached, and of which countries produced what under the division of labour. Plans to economise in energy and in raw material supplies were widely publicised. And last but not least, the point was frequently emphasised that the Soviet Union had agreed to maintain its 1985 level of fuel and essential materials provided the East European countries made reciprocal deliveries of quality foodstuffs, consumer goods, machinery and equipment. Five years after the last Brezhnev Party Congress, Moscow could feel relieved that it had knocked some life back into Comecon, though without achieving the level of integration it wanted. It could also feel assured that it had the final weapon: Eastern Europe would not get Soviet oil if it misbehaved. Yet these were Soviet rather than East European feelings. Away from Moscow there was a sense that the new Comecon thrust was mainly intended for the benefit of the Soviet Union.

Comecon in Soviet policies

In an article in *International Affairs* in December 1985, Yuri Shirayev, a corresponding member of the Soviet Academy of Sciences, laid great stress on the importance of integration in Comecon in the move from extensive to intensive economic development and the promotion of social progress. In the technological age this was essential; and he set out the theoretical argument with great care.[6] At the Comecon Session in the same month, Ryzhkov, the new Prime Minister, spelt this out in political terms. What was necessary was 'the radical intensification of social production' within CMEA:

> The present-day political situation requires the pooling of the efforts of countries of the Socialist Community for effective use of the possibilities opened by the scientific and technical revolution on the basis of the advantages inherent in the Socialist system.

This would help the CMEA itself. But it would also carry the message elsewhere:

> The whole of humanity is comparing and will compare the results of the scientific and technical revolution in Socialist and in capitalist countries. Humanity is already aware that we adopt the programme of peaceful construction for the sake of man. And we are not striving to create a sort of privileged club.[7]

An integrated Comecon would be the best possible advertisement for Soviet-style socialism. It would also, he might have added, be of great help to the Soviet economy.

At the time of the Twenty-sixth Party Congress in 1981 the Soviet economy was floundering badly, particularly agriculture whose miserable performance had forced the Soviet Union to import almost a fifth of its grain. Brezhnev spoke in glowing terms, quoting all the favourable statistics he could find. So did Tikhonov, whose responsibility it was. But tucked away in both their speeches were admissions of failure and admonitions to improve in the 1981–5 Plan. At this point, the reasons for the various shortcomings do not matter.[8] The important thing was to rectify them and to get a sluggish economy moving, and not simply, it might be pointed out, because income was so low, but because expenditure, not least for the military budget, was so high. But at the end of the Five Year Plan the news was not very much better. Writing towards the end of 1985, E. Figurnov, a Soviet economist, could note with pride that 'by the early 1970s, social labour productivity in the USSR had reached 40 per cent of the US level, as compared with Russia's less than 10 per cent in 1913'. But he had to add that 'In the period from 1971 to 1982 ... our economic development somewhat slowed down' and that, in particular, 'social labour productivity growth declined from an average annual rate of 6.8 per cent ... to 2.9 per cent'. Although he could not give figures for the last year of the Plan, he could report, for example, that the average rate of industrial growth between 1981–2 and 1983–4 increased from 3.1 per cent to 4.2 per cent, which was something, but not the rate of increase that the Soviet Union needed.[9] In fact, of course, it declined again in the last year.

Figurnov found a number of 'objective criticisms' to explain this disappointment, from the inaccessibility of raw materials to the vagaries of the climate. But, quoting Gorbachev, he put much of the blame on the country's management. It was, of course, a very difficult period. In his later years, Brezhnev slowed down, and much of the Party and government apparatus with him. Before his final effort to boost the output of agriculture could get into its stride—his food programme—he was dead. Andropov's attempt to instil discipline in management and work-force alike also petered out with his declining health and early death. Chernenko had little mind for the task. Yet at least one of those who aspired to succeed him beavered away at preparing a new scheme for economic growth so that, when Chernenko too died quite soon, Gorbachev was ready for action. And what Figurnov summed up with his economist's pen was Gorbachev's emerging plan: a switch to intensive development through a rapid advance in science, technology and engineering, intended to raise Soviet levels of productivity above those of capitalism before the year 2000, all to be achieved under the direction of a much more efficient form of centralised economic administration. This was the only way to provide for 'a rapid improvement of the people's well-being' and for 'maintaining the country's defence capability'. And on this last point, though Figurnov did not elaborate it, the prospect of possibly having to match SDI must have appeared frightening.

With a new man on the job, the Soviet Union was going to look after the Soviet Union, a perfectly natural attitude. With his technocratic approach, Gorbachev may well have considered seriously the pros and cons of the East European connection. There is now a considerable Western literature on the profit-and-loss account of Eastern Europe to the Soviet Union.[10] On balance this suggests that in the 1970s Eastern Europe received a net subsidy, partly through energy and raw material prices and partly through Soviet military supplies for the Warsaw Treaty Organisation. So it would perhaps hardly have been surprising if Gorbachev had decided to let Eastern Europe go its own way. But there is a certain inertia about long-standing payments. And in present military terms disowning the East Europeans would have been a strategic nonsense, while in foreign policy and ideological terms it would have represented a serious defeat. So, however costly or ungrateful his Comecon partners, he can have seen little choice but to persist in friendship with them. Yet that these thoughts were not simply in Western minds is borne out by what happened to Soviet prices to Eastern Europe for fuel and other essential supplies, and by the blunt views expressed by Ryzhkov. Demands for higher East European contributions to the cost of defence tell the same story. It is clear that, at the end of his calculations, Gorbachev came to the same conclusion as his predecessors, that the subsidy should be greatly reduced.

At the same time, what also emerged was a decision to utilise more effectively what the East Europeans could offer through Comecon, and that was three things: specific commodities, from Hungarian farm-produce to East German machine-tools; capital investment in bilateral and multilateral joint enterprises, particularly in energy and raw materials; and co-operation in science, technology and engineering, where Eastern Europe is sometimes ahead of the Soviet Union. And this decision came via Comecon—or bilaterally—more as a demand than as a request. However, there was possibly a compliment to the East Europeans, or the more progressive among them. Interpreting Gorbachev's grand scheme, Figurnov wrote of a combination of strengthened centralised decision-making in crucial matters, plus broader enterprise independence in other matters. Be that as it might, Comecon certainly had a place in Soviet economic planning. Speaking on a link-up with Swiss television in February 1986, Ryzhkov noted that the integration process with the socialist countries would be accelerated in joint activities planned down to the year 2000. But on another point, he indicated that

> By the year 2000 we intend to double the country's national income ... Very big tasks have been set along all the directions of the country's social and economic development. And, of course, in attaining them we will rely mainly on our own resources.[11]

On the eve of the Twenty-seventh Congress, Soviet leaders knew what they wanted. Eastern Europe might help, and be helped. But naturally and inevitably, the priority was the Soviet Union itself.

Notes

1. *Documents and Resolutions of the Twenty-sixth Congress of the Communist Party of the Soviet Union*, Moscow, Novosti, 1981, pp. 8–40.
2. For a general review of the Twenty-sixth Congress see S. Bialer and T. Gustafson (eds), *Russia at the Crossroads: The 26th Congress of the CPSU*, London, Allen & Unwin, 1982; and for a wide-ranging discussion of the Soviet Union at the time, see A. Brown and M. Kaser (eds), *Soviet Policy for the 1980s*, London, Macmillan, 1982.
3. *Pravda*, 15 and 21 June 1985.
4. For one view of Gorbachev's performance in 1985, see V. V. Kusin, 'Gorbachev and Eastern Europe', *Problems of Communism*, January–February 1986, pp. 39–53.
5. Yu. Mikhailov, 'The Main Potential of Peace', *International Affairs*, no. 1, 1986, pp. 43–50.
6. Yu. Shinayev, 'The CMEA Countries: Dynamic Economic Development', *International Affairs*, no. 12, 1985, pp. 18–24.
7. *Tass*, 17 December 1985.
8. One of several analyses is to be found in M. I. Goldman, *USSR in Crisis: The Failure of an Economic System*, New York, Norton and Co, 1983.
9. E. Figurnov, 'USSR's Growing Economic Might as the Basis of its Effective Foreign Policy', *International Affairs*, no. 11, 1985, pp. 60–7.
10. Cf. S. M. Terry (ed.), *Soviet Policy in Eastern Europe*, New Haven, CT, Yale University Press, 1984, pp. 155–88.
11. *Tass*, 5 February 1986.

4 The Twenty-seventh Party Congress and the future of Comecon

The balance sheet before the Congress

On the eve of the Twenty-seventh Party Congress, several balance sheets could be drawn up for Comecon. The economic one showed extensive intra-area trade, an interesting range of bilateral and multilateral enterprises, a complicated and only partly successful planning system, a disappointingly poor overall industrial and agricultural performance, an inadequate level of applied technology, an insufficient output of consumer durables, and a most ambitious programme for future development. The existing shortcomings raised serious doubts about the prospects for even modest growth. The importance of scientific, technological and engineering advances was recognised, but whether the investment capital, the labour incentives, and the managerial apparatus were wholly adequate was open to question. The fact that much the same could be said about each of the Comecon member countries individually did nothing to quell the doubts.

The fact also that the Soviet economy was the major one and had more than its fair share of the problems was not at all encouraging. On the other hand, with a dynamic new general manager, it was the one with the best chance of achieving its objective of doubling its income by the end of the century. And it might assist the others. Yet its objective looked enormously ambitious; and it might even be held back by the others.

This, of course, was part of the political balance sheet. The Soviet Union had a renewed interest in Comecon, for economic and political reasons. Economically it might gain from the East European link, or find it a continuing financial drain; but, even if the latter, it could still not abandon it politically. And the Soviet view of Eastern Europe was a mixture of hope and dread, in political as well as economic terms. Eastern Europe might maintain the Soviet model of socialism, be a useful advertisement and a dependable ally; or on recent evidence it might be a growing embarrassment. Yet at the end of the day, the Soviet Union was its own master, and it had other things to think about, too.

Conversely, Eastern Europe did not have charge of its own destiny, and it had the Soviet Union as its major preoccupation. Gorbachev's management might lead the entire Comecon economy into a period of plenty. Alternatively, however, it might simply thwart East European initiatives and lead to worse terms of trade within the Soviet bloc. Equally, Gorbachev might reach a military accommodation with the United States and an economic relationship with the Common Market, so that Eastern Europe would have some freedom of movement in its dealings with the outside world. Yet for reasons of internal opposition or external hostility, he might perhaps fail, in which case Eastern Europe would be drawn more closely into a Soviet-dominated, autarchic Comecon. Against this background, Eastern Europe was cautious. There was almost universal official support for Gorbachev's reported intentions, and also widesprad public interest. But at all levels the barriers had been strengthened and were being kept fairly intact. Up to a point, the Soviet Union could ignore Eastern Europe; but Eastern Europe had an inferior alternative—to tend to its defences against the Soviet Union. So the political balance sheet was reasonable if the economic balance sheet was promising. But otherwise not. And the East European barriers might themselves undermine economic progress.

There was also a military balance sheet, but that essentially depended on progress in East–West arms talks. Successful talks could mean an improvement in East–West trade (and, of course, vice versa). And the majority Comecon view now favoured an increase in trade. The Hungarians had been campaigning for it against Soviet expert opinion; at last, in the right conditions, Ryzhkov advocated it. As he indicated to his Swiss audience,

> to utilise the possibilities inherent in the development of East–West economic ties enters into a profound contradiction with all sorts of discriminations, embargoes, trade and credit restrictions that are being extensively practised today by the governments of the United States and some other Western states ... In practical terms the Cocom instructions mean that only outdated products and technologies can be sold to Socialist countries. But we will never accept such a so-to-speak 'division of labour'.

East–West trade should not amount to a war. There could and should be trade based on the 'principles of full equality, mutual advantage, strict observance of existing accords, and non-interference in internal affairs':

> We see possibilities for a considerable development of our external economic ties ... The pace of the Soviet economy's development, its scale, our consistent course of developing international economic co-operation allow for the conclusion that the prospects of business co-operation with the Soviet Union ... can be very promising.

Ryzhkov was equally in favour of investment:

> We will not be against drawing Western firms or credits into the fulfilment of our investment projects. The countries and firms that will be ready to base their economic relations with the USSR on mutual advantage and equality will always find the Soviet Union a reliable and serious partner.

He also advised Western firms to make a close study of the Five Year Plan that had just started; and he specifically suggested participation in the 'reconstruction and modernisation of Soviet enterprises'.[1] His speech was Soviet-orientated; but the principles and proposals applied equally to Comecon as a whole. Given renewed—or genuine—*détente*, the Soviet Union and Eastern Europe could together modernise through trade with the West, including through borrowing. On that basis there might be rather more hope for their economic development, both separately and in harness.

As the Congress approached, therefore, there was an air of mixed anticipation and apprehension. It was clearly an important occasion for the Soviet Union, its leaders and its people, regardless of Comecon's future. Despite the fact that it was not their Congress, it was also important for the East European peoples, and no less for their leaders. Of the latter, five were already of the old generation of septuagenarians discarded in the Soviet Union; and it seemed significant that in mid-February the least flexible of them, Husák, hinted at possible reforms. But whatever they did, the end of their time was approaching. Four local party congresses also lay ahead: the Czechoslovak in March, the Bulgarian and East German in April, and the Polish in June. And after his own show, Gorbachev would be powerfully placed both to influence East European party congresses and to determine the succession where necessary.

Gorbachev, the Party and Comecon

At the end of the Twenty-seventh Congress Gorbachev was certainly more entrenched in power and more enhanced in authority. Although what may have been his own campaign to change the rules and so weaken the Party bureaucracy had had no apparent success, and although at least two of the old guard managed to hang on to their positions in the Politburo, there was a two-fifths turnover in the Central Committee. And although his economic programme lacked precision, his marathon introductory speech was both robust and full of promises.[2] Naming names, he attacked both corruption and inefficiency. He condemned alcoholism and praised old family values; he called for better discipline and higher productivity. But he went on to offer a higher standard of living through a complex of measures, including improved central planning, a degree of autonomy to some industrial enterprises, the introduction of marketing to a proportion of collectives' production, and the widespread application of science and technology throughout the economy. What will happen in practice is another matter—whether productivity will rise enough in ten to fifteen years to satisfy consumer demand at the same time as meeting investment requirements. But Brezhnev (though in this case not named) had been a failure. According to Gorbachev,

> Difficulties began to build up in the economy in the 1970s, with the rates of economic growth declining visibly. As a result, the targets for economic development set in the CPSU programme, and even the lower targets of the 9th and 10th Five Year Plans, were not attained. Neither did we manage to carry out fully the social programme charted for this period. A lag ensued in the material base of science and education, health protection, culture and every day services.

Following the Party Congress, one objective would be to remedy these shortcomings. But the over-riding objective would be the 'acceleration of the country's socio-economic development' since that 'is the key to all our problems: immediate and long-term, economic and social, political and ideological, internal and external ... the only way a new qualitative condition of Soviet society can and must be achieved'. In short, Gorbachev made it abundantly clear that the future of the Soviet Union depends on its economic success, and that, whatever it might take, he was committed to it.

In this the CMEA had a role to play. As Gorbachev elaborated:

> We are convinced that Socialism can resolve the most difficult problems confronting it. Of vital significance for this is the increasingly vigorous interaction whose effect is not merely the adding up but the multiplication of our potentials and which serves as a stimulus

for common advancement. This is mirrored also in joint documents of countries of the Socialist Community.

Interaction between governing Communist Parties remains the heart and soul of the political co-operation among these countries. During the past year there have practically been no fraternal countries with whose leaders we have not had meetings and detailed talks. The forms of such co-operation are themselves being updated. A new and perhaps key element, the multilateral working meetings of leaders of fraternal countries, is being institutionalised. These allow for friendly consultations on the entire spectrum of problems of Socialist construction, on its internal and external aspects.

It had been a busy year, and there were even busier times ahead.

Just what Gorbachev had in mind in the economic sphere he did not spell out in any detail; but it would involve a shake-up, and his phraseology about the CMEA echoed his language concerning the USSR:

There is now the comprehensive programme of scientific and technological progress. Its import lies in the transition of the CMEA countries to a co-ordinated policy in science and technology. The accent is being shifted from primarily commercial relations to specialisation and co-operation of production, particularly in heavy engineering. In our view, changes are also required in the work of headquarters of Socialist integration—the Council for Mutual Economic Assistance. But the main thing is that in carrying out this programme there be less armchair administration and fewer committees and commissions of all sorts, that more attention be given to economic levers, initiative, and Socialist enterprise, and that work collectives be drawn into this process. This would indeed be a committed approach to such an extraordinary undertaking.

Vitality, efficiency, and initiative—all these qualities meet with the imperatives of the times, and we shall strive to have them spread throughout the system of relations between fraternal Parties.

Whatever the new approaches and the new energy characterising Soviet economic policy, the East European states will be affected by it.

However, in Gorbachev's view, this would not be a one-sided process:

The CPSU attaches growing significance to live and broad communication between citizens of Socialist countries, between people of different professions and different generations. This is a source of mutual intellectual enrichment, a channel for exchanges of views, ideas, and the experience of Socialist construction. Today it is especially important to analyse the character of the Socialist way of life and understand the processes of perfecting democracy, management methods and personnel policy on the basis of the development of several countries rather than of one country. A considerate and respectful attitude to each other's experience and the employment of this experience in practice are a huge potential of the Socialist world.

Some might argue plausibly that Gorbachev's programme contained tiny elements of Hungarian economic devolution. And others noted at the time that Jaruzelski was particularly well treated at the Congress. Gorbachev's speech had its soothing as well as its abrasive elements for CMEA members from Eastern Europe.

Yet ultimately, Gorbachev's speech was only the words of one person. And again, whether what he put in words will be put into practice is another question. Inevitably the issue was not just a matter of economics. As he saw it:

> The destinies of peace and social progress are now linked more closely than ever before with the dynamic character of the Socialist world system's economic and political development. The need for this dynamism is dictated by concern for the welfare of the peoples. But for the Socialist world it is necessary also from the standpoint of counteraction to the military threat. Lastly, in this lies a demonstration of the potentialities of the Socialist way of life. We are watched by both friends and foes. We are watched by the huge heterogeneous world of developing nations. It is looking for its choice, for its road, and what this choice is will depend to a large extent on Socialism's successes, on the credibility of its answers to the challenges of the time.

In other words, the maintenance of defence and the expansion of influence were at least as important as economic growth, or certainly from Gorbachev's point of view. No more now than before is CMEA business purely economic.

The same theme of defence and infuence was taken up in the Party Programme.[3] Of course,

> In the field of economic relations, the CPSU favours the further deepening of Socialist economic integration as the material basis of the cohesion of the Socialist countries. It considers it especially important consistently to pool the efforts of the fraternal countries in

key areas of the intensification of production and the acceleration of scientific and technical progress in order jointly to accomplish a task of historic significance—coming to the forefront of science and technology, with a view to further growth in the well-being of their peoples and the strengthening of their security.

The Party proceeds from the premise that integration is called upon, to an increasing extent, to facilitate the progress of social production and the Socialist way of life in Community countries, the acceleration of the process of equalising the levels of economic development, and the strengthening of Socialism's position in the world.

But defence was crucial:

The Soviet Communists favour the more effective interaction of the fraternal countries in the international arena, taking into account the situation and interests of each of them and the common interests of the Community.

In conditions of the continued existence of the NATO imperialist military bloc, the Party considers it necessary to promote in every way the improvement of the activity of the Warsaw Treaty Organisation as an instrument of collective defense against the aggressive aspirations of imperialism and of the joint struggle for a lasting peace and the expansion of international co-operation.

And the extension of influence was never far away:

The Party will continue to promote the strengthening of an awareness of the unity and common historical destiny of the fraternal peoples. Disseminating the truth about Socialism, unmasking imperialist policies and propaganda, rebuffing anti-Communism and anti-Sovietism, combating dogmatic and revisionist views—all these tasks are accomplished more successfully when Communists act in a united front.

The CPSU considers it its internationalist duty, together with other fraternal parties, to strengthen the unity and increase the might and influence of the Socialist Community. The course of the competition between Socialism and capitalism and the future of the world civilisation depend to an enormous extent on the strength of this Community, on the success of each country's constructive activity and on the purposefulness and co-ordination of their actions.

In effect, the Soviet Party Programme merely elaborated the principles that Gorbachev encapsulated in a single paragraph. Yet, in three further paragraphs, it took a much less accommodating attitude on relations between Socialist countries, extrapolating perhaps where Gorbachev was still insufficiently clear in his own mind:

> The experience of the CPSU and of all world Socialism indicates that the most important factors in its successful progressive movement are the fidelity of the ruling Communist and Workers' Parties to the teaching of Marxism–Leninism, the creative application of this teaching, the Parties' strong ties with the broad masses of working people, the strengthening of their prestige and leadership role in society, the strict observance of Leninist norms of party and state life, and the development of Socialist people's rule; sober consideration of the actual situation, and the timely and scientifically substantiated solution of problems that arise; and the building of relations with other fraternal countries on the principles of Socialist internationalism.
>
> Whatever the special features of each Socialist country—its economic level, its size and its historical and national traditions—may be, they all have the same class interests. That which unites the Socialist countries is the most important thing and is immeasurably greater than that which may divide them.
>
> The CPSU is convinced that the Socialist countries, fully observing equality and showing mutual respect for national interests, will continue to follow the path of ever greater mutual understanding and convergence. The Party will promote this historically progressive process.

And in two other paragraphs it laid down the law in a wholly unambiguous way:

> The Party seeks durable, comradely relations and many-sided co-operation between the USSR and all the states of the world Socialist system. It proceeds from the premise that the solidarity of the Socialist countries corresponds to the interests of each of them and to their common interests, and that it serves the cause of peace and the triumph of Socialist ideals.
>
> The strengthening in every way of the friendship between the Soviet Union and the countries of the Socialist Community and the development and improvement of the ties between them is a subject of special concern to the CPSU.

Brezhnev was dead; and Gorbachev had in fact buried him without a tombstone. But the Brezhnev Doctrine is still alive, enshrined in the Party Programme. And while General Secretaries come and go, the

Party goes on for ever—a fact Eastern Europe will no doubt have cause to reflect upon.

The future balance sheet

If Gorbachev favoured anyone in the run-up to the Party Congress, it was Husák, the leader at that time least likely to fall out of line. If he cold-shouldered anyone, it was the maverick Ceauşescu. He was cautious about Kádár's domestic and foreign policy initiatives, anxious that Zhivkov should get his economy right, and worried at the people Honecker chose to talk to. Yet it was Jaruzelski who was emerging as the prototype of the new East European leader, sensible, effective and rather younger.

It was also Jaruzelski who was most conspicuous during the Party Congress and most fêted after it. Since then, the Czechoslovaks, the Bulgarians, the East Germans, and finally, in June 1986, the Poles have all held their own Party Congresses and reacted to Moscow's views. Husák has moved a little. Zhivkov has at least given the appearance of being anxious to please, though not changing very much. Honecker has achieved a moderate improvement in his relations with Gorbachev, while keeping his powder dry. Jaruzelski has received Gorbachev's blessing as the most loyal, purposeful and likeable East European leader, but he has also maintained a peculiarly Polish character to his post-Solidarity settlement. Yet whatever happens to individual leaders or to particular states, Eastern Europe will inevitably raise difficulties for Gorbachev, and specifically for Comecon, for some time to come. The barriers have not been torn down; and Gorbachev has only stated that they must be, and suggested possible means ultimately towards that end. His Party colleagues have also made noises likely to keep them up.

His grandiose plan to catch up and overtake the West through better management and more advanced technology is somewhat short on published detail but might perhaps work. An initial success and some further elaboration could start a continuing cycle of development. The economic arguments are set out in later chapters. Criticism tends to focus on his reliance on yet more but better planning. But Ryzhkov has not been the only one to talk firmly about 'direct production ties between Soviet enterprises and enterprises in CMEA member states', a proposal which, if implemented, though it would take a long time, would go part of the way towards meeting the Hungarian criticism of the lack of micro-level connections.[4] And the introduction of quotas for collectives in the Soviet Union itself has a vaguely Chinese flavour to it. So, in the long term, there may be a trend towards successful market socialism within Comecon as a whole.

However, even if the opposite happened and Comecon's planners tried to isolate themselves from the world market, they would be unlikely to have much success. They need advanced Western know-how, and to get it they need to sell energy and raw materials. At the time of writing, the world price of oil is low; and this could seriously jeopardise any Comecon modernisation, starting with the USSR itself. CMEA is not an island. But for that very reason it cannot escape the influence of beneficial as well as harmful world forces, and any setback from oil would be only passing.

Nor can economics be separated from politics. It is not only Eastern Europe that may cause Gorbachev difficulties; so may the Soviet Union itself. Bureaucrats and managers may not take easily to demotion, or generals to reduced funding, or simple workers to redeployment. In some ways the Soviet Union is an immobile society, which is the opposite of what Gorbachev needs. He is also young only by the standards of the immediate past, and there are bound to be younger men with ambitions to take his place. So the Comecon plans of today may not be those of tomorrow, which might produce a yet different response from Eastern Europe. It is possible, but not too easy, to build up an optimistic economic scenario. To build up an optimistic political one takes a little more effort still.

It is an interesting theory that most East European problems stem from the efforts of monocentric governments to control their pluralistic societies. The theory could be extended to the USSR and, in a doubly complex fashion, to the CMEA—the Soviet Party attempting to run all the other communist parties, and each in turn trying to run its own country. Viewed in this way, the streamlining and modernisation of Comecon is a non-starter, and certainly its integration. This is one of the sharpest contrasts with the Common Market; and yet even there, with no monocentrism, there have been problems. Much will depend, therefore, on whether Gorbachev inclines to listening to East European as well as Soviet views, and to Soviet views additional to those of the hierarchy.

The prospects for Comecon turn also on international politics. A renewed arms race would quickly skew technological development, damage consumer prospects, undermine productivity. and intensify centralist tendencies. There would be no general, willing modernisation either in Comecon or in its constituent parts. A semi-arms race would have lesser effect. But the Gorbachev assumption is of renewed *détente* which, on current showing, is not unreasonable, remembering that the previous *détente* was just lowered tension. In such a situation, Gorbachev would have the incentive to make Comecon successful. It stands to reason that he wants it to rival the Common Market, just as he wants the Soviet Union to rival the United States. This would be advantageous not only for equilibrium in Europe and for balance against America (some would argue, wrongly, superiority). It would increase the likelihood of other states wishing to be associated with Comecon (as many are with the Common Market), and the possibility of their being accommodated without overloading the joint economy and arousing the deep suspicions of the East European states.

To be attractive in this way, of course, Comecon would have to be not only economically prosperous, but also politically welcoming; and at present it is neither, except to those countries which are desperate even for low-level assistance, or too far away or too alike to be concerned about the character of internal government in Comecon. In Europe, at the moment, the effect would be the other way round—which is one of the reasons why Gorbachev is cautious in approaching EEC, no matter how much he would like its support in technology and arms negotiations. However, even a marginal improvement in CMEA performance would help the Soviet Union in its Third World aid programmes, which currently trail behind those of the West.

In a sense, too, the Third World, broadly defined, may be the main Gorbachev objective in trying to revitalise Comecon.[5] At the Party Congress he called for a 'bold and creative approach to the new realities' of the developing world, though he did not spell out what it would be. But the existing Soviet economic model has lost its shine; and Comecon's armchair administrators have done nothing for their image. Yet, in less than a decade, China has produced an economic model that is relevant and highly successful. So, before it also enters the CMEA-type field, it is incumbent on Gorbachev to breath life into both the Soviet economy and Comecon. That might even involve learning something from the Chinese, which could do quite a bit of good, as well as acquiring technology from the West. The Soviet Union and Eastern Europe are so well-endowed and their peoples are so talented that it would be sad if they did not get the opportunity to transform Comecon and make it a constructive influence in the contemporary world.

Notes

1. *Tass*, 5 February 1986.
2. *Novosti*, 25 February 1980.
3. *Current Digest of the Soviet Press*, XXXVII, 44, 27, November 1985.
4. *Tass*, 5 February 1986; cf. M. Racz, 'Inter Country Relations at the Microeconomic Level: The Intra-CMEA and the Hungarian Experience', *Trends in World Economy*, no. 51, Budapest, 1985.
5. On the problems of CMEA and the Third World, see C. Coker, *The Soviet Union, Eastern Europe and the New International Economic Order*, Washington, DC, Sage, 1984.

5 Comecon and Common Market

The machinery of Comecon

One of Comecon's most frequently cited handicaps is its actual machinery, which has evolved little since Brezhnev's death.[1] At the centre is the so-called Council Session (see Figure 5.1). This normally meets annually and comprises government delegations, which are usually led by a permanent representative of the standing of a deputy prime minister but may occasionally be led by someone more senior. In appearance, therefore, it is a high-level body with a steady remit. But the reality is a little different. In the first place, within the Soviet and East European system, government ministers rank below Communist Party officials. The Council Session is thus at most a second-level body. This subservience is underlined by the regularity with which Sessions are preceded or accompanied by conferences of the Party secretaries of Comecon members. Agreements produced by the Council Session tend to be merely elaborations of joint Party directives, a point that is borne out by the new Soviet Party Programme. Since 1984, however, Sessions have been held twice a year with much more senior representation and more effective deliberation. Yet, in the second place, a unanimity rule still persists. It is now open to any member to declare its 'lack of interest' and therefore to enable the remainder to reach agreement. Yet even where members' interests may appear to be slight, they are not always willing to admit it and, on crucial matters, they retain their disabling veto. Thirdly, on other than organisational matters, agreements reached by the Council Session are recommendations, not decisions; they still have to be accepted collectively by the individual governments and executed through bilateral or multilateral agreements. In many ways, the Council Session appears to be little more than a two-way channel of communication.

In between meetings, however, there is an Executive Committee, made up of the permanent representatives, that convenes at least four times a year in Moscow. The least important of its functions is its major one, to process the various agreements reached by the Council Session.

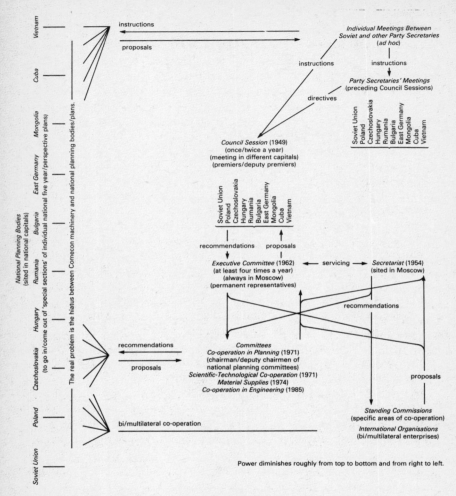

Figure 5.1. Simplified diagram of CMEA decision-making procedure

Gradually it has developed the more useful purpose of supervising work on plan co-ordination and scientific and technological co-operation, and of receiving the reports of most of the Council's subsidiary bodies. Nevertheless it has not succeeded in acquiring the role that Khrushchev intended when he founded it in 1962, that of supranational planning committee. And it has failed despite the steady growth of a Council secretariat, housed in Moscow and headed by a prestigious Soviet official.

The body that gets nearest to a proper planning organ is the Committee for Co-operation in Planning, comprising the chairmen or deputy chairmen of the various national planning offices. As one of only four committees, as distinct from commissions, it has standing as well as function. Yet despite its access both to the Executive Committee and to national bureaucracies, it too remains international rather than supranational through lacking decision-making powers. Yet with its two fellow Committees on Scientific-Technological Co-operation and on Material Supplies, and the new Committee on Co-operation in Engineering formed in 1985, it certainly adds to the weight of machinery at the centre pressing for genuine integrative activities in the spirit of the Complex Programme.

Lower in the framework, a whole batch of standing commissions cover almost every aspect of economic activity, from agriculture to atomic energy, and do the donkey-work for co-operation in specific areas. There are also periodic conferences and well-staffed subordinate institutes. All of these contribute their might to co-ordination; but they are even less powerful than the committees they report to. More recently, various so-called international organisations have taken on specific duties at a decentralised level with some success; and it could be argued that, with their limited objectives and specialised personnel, they offer the best hope yet of effective integrative machinery, however, low in the Comecon hierarchy they may be.

What lends significance to the vexed question of Comecon machinery is the impossibility of integration other than by decisions at the centre. The economic debates of the 1960s envisaged at one stage the decentralisation of Comecon to the point where commodity convertibility would be more important than plan co-ordination. This is a view still held by many Hungarian and other economists; and under such a system, many of the changes in an individual country's economic output that now necessitate complicated negotiations—if indeed they get anywhere at all—could have been achieved by automatic means. But in practice the whole trend since 1968 has been the other way. Even the so-called convertible rouble is not really convertible. Yet a major problem with plan co-ordination is the sheer complexity of the task. In the early days it was simply a matter of trading with production surpluses. Gradually effort was put into attempting to co-ordinate planned investment and production. But there remained the problem of many countries and many products; and this has only been partly solved. The plethora of committees and organisations established in the 1970s is still unable to grasp all the economic intricacies of the Soviet Union and Eastern Europe, despite the fact that their scope has extended further than some governments wanted. There are many subsidiary reasons for this, not least the lack of a good database and of adequate computer coverage. But one of the main reasons is the plain impossibility of the task. There is simply no kind of machinery that can cope with the multiplicity of interlocking decisions that have to be made more or less simultaneously at different levels and involving conflicting interests. And the more complex individual economies become, the more unlikely it appears that a machinery will emerge.

Yet in some ways it might be said that a successful war of attrition is being fought against the unmanageability of central Comecon control. The Committee for Co-operation in Planning has promoted discussions on both the medium and long term and has succeeded in implanting bloc proposals in national plans at the formulation stage, leaving their execution to subsequent bilateral or multilateral agreements. Thus, in the period 1976–80, all Comecon countries, except Romania, at least introduced so-called 'special sections' to their five-year plans, setting out the national resources devoted to agreed international Comecon projects. Similar 'special sections' now appear also in perspective plans, committing the various national economies in principle to supplying the investment necessary for longer-term proposals; and these, too, are being co-ordinated through the Committee for Planning in the form of the long-term Comecon Target Programmes previously referred to. The Soviet establishment in particular appears to be optimistic about the growing percentage of national activities likely to be pressed towards co-operative enterprises in this way. In addition, the various forms of international organisation since established to supervise the co-operative enterprises on the ground are, as already suggested, promising examples of practical collaboration at local level, once the central planning hurdle has been passed. And this is one of the activities that Gorbachev gave a boost to in his Twenty-seventh Congress oration.

All this might be termed *ad hoc* co-ordinated planning or piecemeal integration. Soviet propagandists have been vociferous in praise of specific achievements. The list includes: the Mir power grid, the Druzhba oil pipeline, and the Soyuz gas pipeline; the Yermakovo ferro-alloy and the Ust-Ilmisk cellulose plants in the USSR; and the Katowice iron and steel combine in Poland, the Devna soda factory in Bulgaria, and the Böhlen ethylene complex in East Germany. It is an impressive array; and there is little doubt that many more such co-operative enterprises will be discussed in the near future. There is, after all, a particular incentive, which is not devotion to Comecon, but economic need. Despite the rich natural resources with which the Soviet Union in particular is endowed, Comecon as a whole faces a shortage of energy, raw materials and chemicals. Present methods of production are more often than not inefficient. Existing supplies are frequently misused. Imports are costly, and exports could earn valuable hard currency. In many instances, too, the investment required for expensive operations would be beyond the immediate capacity of any one country, even of the Soviet Union itself, given the other claims upon it. So co-ordinated activity of this kind is high on the Comecon agenda.

However, even the most enthusiastic Soviet economists in Brezhnev's time expected integration of a far-reaching kind to take a long time. 'The first stage', wrote Yuri V. Shishkov of the Institute of World Economy in 1979, 'started with the adoption of the Comprehensive Programme . . . in 1971 and will last for approximately two decades.'[2] And at least two further stages were anticipated. In addition, Soviet economists rationalised piecemeal integration. It was taken, in effect, as one of a 'number of "partial integrations"—the integration of the national commodity markets, of the national markets of capital, of the economic policies of the respective governments, and so on', that, taken together, would eventually produce a centrally planned Comecon.

It may be that in the course of time—two, three or four decades from now—this is what will happen. The machinery of Comecon may evolve to take account of multiplying planning successes until the Council Session commands all the individual economies as one. But many political and economic questions have been raised and very few settled. And meantime practical organisational problems remain. Looking ahead a mere decade and a half in October 1985, *International Affairs* commented on what is required by way of a 'complex of interconnected economic and organisational measures both within the framework of individual countries and in the CMEA's administrative system'. It did not make the objectives seem easy to manage—

> to co-ordinate economic policies, promote co-operation in the field of planning, develop mutual trade and perfect commodity–money instruments, bring the structure of the national economic mechanisms closer together in a purposeful way, and strengthen the organisational and legal foundations of the CMEA's activity.[3]

All of which sounds like more, not less, of the 'armchair administration' that Gorbachev attacked at the Party Congress.

Comecon and Common Market: some comparisons

The attitude of the Soviet Union to Comecon is one thing. Its attitude to the Common Market is another.[4] This has in the past been somewhat paradoxical, but it appears to be changing. The Soviet Union has been unwilling to recognise the existence of the Common Market and certainly not its successes; but it has also been quite willing to buy its cheap surplus meat and butter. It has opposed individual relationships between the Common Market and East European countries and has insisted that any trade agreement should be on a group-to-group basis. In the early summer of 1985, however, it dropped its insistence on this last point and suggested that a joint declaration on trade would suffice. It also went so far as to propose that this would be an adequate platform for direct trade agreements between the Common Market and the various members of Comecon.

When the Common Market turned down a previous approach, it did so on the ground that Comecon had no responsibility for its members' foreign trade; and although it is showing interest in the more recent approach, it is sending individual replies. Exactly what are the motives of the two sides is neither clear nor relevant at this point. But the issue of responsibility for foreign trade is only one of a number that differentiate the two organisations and make their working together very difficult. There are many others.[5]

The first difference between them lies in their general concept. The Common Market emerged in 1958, after the worst of the Cold War had passed away, as a voluntary association of six fairly like-minded and roughly equivalent states that wished to increase trade among themselves by removing internal customs and erecting an external tariff barrier. They also agreed to have a budget for specific economic and social purposes and to pursue common policies in crucial areas such as agriculture and fisheries. In this sense, therefore, they decided of their own free will to share a certain amount of their individual sovereignty. This is also true of the six members who have since joined. To that extent, the EEC might be described as a supranational organisation. At the same time, all its members have made it clear that they could withdraw from membership and that there are limits to their surrender of power. In particular, though they allow more scope to their shared institutions now than they once did, they have by and large resisted supranational planning. Each has sovereignty over its own development, within the general rules. In addition, there are large areas where private enterprise, national and multinational, operates quite free of any serious kind of local or central planning.

The origins of CMEA were quite different. It was set up during the Cold War as a result of the pressure of a very large state upon a group of smaller ones; and it brought together a series of locally planned economies with only the eventual notion of forging a single centrally planned economy. In one sense, all the small states surrendered their sovereignty completely, and they have several times been forcibly reminded of this through Soviet military action. On the other hand, Albania was able successfully to drift away by ceasing to attend meetings after 1961; and Cuba and Vietnam, not only as late joiners, but also as distant members, have surrendered little of their sovereignty. However, this does more to illustrate the true character of the CMEA, which is the dominance of the Soviet Union over the core East European members. In this sense, the CMEA is a genuinely supranational organisation. Yet this is a distorted and one-sided sense; it is political, not economic, for CMEA members have not traded in much of their economic sovereignty.

Each Comecon state determines its own internal economic policy. In theory they all share the same economic goals and have identical social and political systems. But their levels of economic development are varied, and their political cultures differ considerably. So their policies are directed to disparate objectives and are greatly influenced by local conditions. But above all, each state prepares its own five-year plan, which starts from its own aims and needs, which does not from the outset incorporate common policies, and which only gradually and marginally adapts to Comecon interests as expressed in co-ordinating meetings. The effects on the various domestic economies are not very marked. Co-ordination of plans extends mainly to end-products and seldom reaches down to sub-assemblies. There is certainly nothing equivalent to the interventionist impact of EEC common policies. And specifically in agriculture, for example, there is no specialisation and thus no support for and no restrictions on particular crops grown by individual CMEA members.

On top of that, economic relations among the different Comecon countries are predominantly on a bilateral, not a multilateral basis. Trade is by quota, both for products and for countries. This is overwhelmingly the case, too, with joint enterprises; and where they are multilateral, they never embrace all Comecon members. There are no compensating multinational companies of the kind that are so common in Western Europe. And there is not one single supranational economic integrating organisation of the kind of the European Coal and Steel Community in the EEC.

Just as the internal trade is mainly bilateral, there is also no common tariff against the outside world, and foreign trade is conducted on the basis of individual country-to-country agreements. National protectionism is standard practice. There is no free movement of currency among Comecon members. As a unit of accounting the transferable rouble has a restricted money function. Where world prices are used as a measure, the relation between them and individual currencies is once again determined by the state or states and not by some kind of free flow as in Western markets. Capital moves within Comecon only with the agreement of governments and therefore in the interests of individual states. There is no buying or owning of other countries' assets. The migration of labour is restricted by governments devoted to maintaining full employment at home as a point of socialist honour, or anxious to avoid shortages of labour where machinery is short or demographic profiles uneven.

There is thus a paradox within Comecon. Politically it is supranational; economically it is largely autarchical. The East European states have surrendered the independence of their political assemblies but they hold on to their cabbage patches. In the Common Market it is somewhat the other way round; parliaments are supreme, but agriculture takes orders.

Inevitably, of course, that is a distortion. In the EEC, individual parliaments influence or sometimes determine common policies: thus the British help to shape the common agricultural policy, whereas the French and Germans more or less established it and still dominate it. Equally, the Commission uses its initiatory and executive functions to sustain a leading role for the supranational bureaucracy; but it is still accountable to the Council and to meetings of ministers, and it is increasingly challenged by the European Parliament. Similarly, East European assemblies are adopting a slightly higher profile in the context of more sophisticated, pluralistic societies; and leaders from Warsaw, Budapest and Berlin are influencing CMEA and WTO policies at consultative meetings or on individual visits to the Soviet Union. In addition, through what it supplies and what it charges for oil and raw materials, if in no other way, the Soviet Union has a lever with which it can influence the main domestic lines of East European economic policies. Yet the essential truth remains that members of Comecon, in comparison with Common Market countries, are dominated by a single power politically but behave economically in a remarkably autonomous fashion. The fact of the matter is that, not long after their political independence was taken away, their economic independence was both strengthened and enshrined.

Equally, too, CMEA lacks the unifying factor present in the EEC, that is, transnational market forces. There is some devolution to joint enterprises and the international organisations that run them, and there are single enterprises that seek Comecon outlets as well as trade with the West or the Third World. But the barriers to the movement of the factors of production force simple decisions upwards to government planners, who think in terms of the national product and have no wish to surrender their suzerainty—down, or up. Strengthened by the requests coming from below, they see no need to acquiesce in proposals emanating from above. And so Moscow, despite, or perhaps because of its overweening political power, finds itself economically frustrated as it tries to modernise Comecon for the twenty-first century.

Yet the CMEA apparently has a somewhat similar administrative structure to the EEC. Parallels can be drawn between the Council Session and the Council of Ministers, and between the Executive Committee and the Commission. But these are more apparent than real, since the Council Session and the Executive Committee lack the interventionist powers of their opposite numbers. In the past few years, too, there have been changes in both structures.

The EEC Council of Ministers continues to function as a decision-making body with a ministerial membership varying according to the specific issues involved, and it gets things done. But in 1975 the European Council was established as a formalised meeting of government heads intended to expedite and democratise supranational planning.[6] Its impact has been considerable; but recently it reviewed the entire operational structure of the EEC and agreed to empower the Council of Ministers to work towards completion of a single internal market by 1992, using majority decisions rather than awaiting unlikely unanimity to force things through. The decisions of this Luxemburg meeting of the European Council have to be ratified by individual governments, and there is a broad spectrum of views between those who favour and those who distrust a complete customs union and more supranational decision-making. On the other hand, there is now a very much stronger push at the top towards a genuine European Community; and the powers of the European Parliament may also be strengthened, which would increase the push from below. In other words, the pressure in the Common Market is now firmly towards more supranational activity based on responsible government involvement and developing democratic participation.

There has also been a movement on the CMEA side. For the past two years, as already indicated, the Council Session has met twice instead of once a year; and the level of participation has risen to that mainly of prime ministers. As with the European Council, too, collective statements have been made on important issues of international affairs. Yet there is no CMEA body corresponding to the European Parliament; and the Council Session, even at an elevated level, is in no way a meeting of equals. Whatever the machinery, the fact remains that the Soviet government is the dominant political force. In addition, there is another machinery external to the CMEA organisation and detracting from its importance. This is not so much the WTO, although that is coterminous with the European CMEA in a vital way in which NATO is not with EEC. It is rather the inter-Party nexus, which removes a great deal of decision-making from inter-governmental groups and tends to act as a transmission belt for Soviet views. While all of this strengthens the Soviet Union's political hold over its fellow members of CMEA in Eastern Europe, it reinforces their determination at government level to look after their own economic interests; and it does nothing to win popular support, let alone to soothe popular distrust.

None of this is meant to suggest that everything is rosy in the Common Market. There are many problems, from the Common Agricultural Policy to interminable budgetary disputes. There are serious economic difficulties confronting both individual members and the entire organisation. But there is measurable progress towards integration and a new spirit behind it. And the relations between the different societies are multifaceted and capable of much further development in the future.

By contrast, while Comecon countries both collectively and individually have many successes to report, and while their production difficulties are not entirely dissimilar, at least in their origins, to some of those experienced in the West, they are still not committed to integration; and indeed their actions often seem designed rather to thwart co-operation and to promote protectionism. As a result, they appear to be generally less successful than their neighbours in dealing with the run-up to the year 2000. All the ills of East European agriculture, of course, cannot be placed at the door of non-integration; nor for that matter the great lack of high technology. But in an area where the specialists talk quite freely about the economics of shortage,[7] it is not without significance that their political leaders put enormous emphasis—at least in their speeches—on the urgent need to exploit the international division of labour through CMEA. Whether such a division can be achieved, of course, without a major change in the politics of Comecon as a whole and of its member countries is a separate question. Certainly, if there is no political shift, the economic prospects for Comecon, compared with the Common Market, are not good. Against this background it is difficult to hold out very great hopes for Gorbachev's dream of a co-ordinated Comecon technological revolution, despite his own apparent enthusiasm and that of quite a number of East Europeans.

Notes

1. For a detailed account of Comecon organisation, see G. Schiavone, *The Institutions of Comecon*, London, Macmillan, 1981; and O. A. Chukanov (ed.), *Khozjaistvenni mekhanizm v stranakh-chlenakh SEV*, Moscow, Politizdat, 1984. Cf. Also A. H. Smith, 'Plan Coordination and Joint Planning in CMEA', and Yu. V. Shishkov, 'New Research into European Integration in the Soviet Union', *Journal of Common Market Studies*, vol. 18, no. 1, 1979, pp. 3–21 and 97–113.
2. Shishkov, *op. cit.*
3. V. Kirpichnikov, 'On the Results of the CMEA's 40th Session', *International Affairs*, no. 10, 1985, pp. 32–7.
4. Yu. Shishkov, 'The EEC in a Vicious Circle of Problems', *International Affairs*, no. 11, 1985, pp. 64–73.
5. For useful discussions of relations between CMEA and EEC and their different integrational approaches, see J. Pinder, 'Integration in Western and Eastern Europe: Relations between the EC and CMEA', *Journal of Common Market Studies*, vol. 18, no. 1, 1979, pp. 114–33; G. N. Yannopoulos, 'EC External Commercial Policies and East–West Trade in Europe', *Journal of Common Market Studies*, vol. 24, no. 1, 1985, pp. 21–38; F. Kozyma, *Economic Integration and Economic Strategy*, Budapest, Akadémiai kiadó, 1982, pp. 105–26; G. Schiavone (ed.), *East–West Relations: Prospects for the 1980s*, London, Macmillan, 1982, pp. 133–69; and A. Köves, *The CMEA Countries in the World Economy: Turning Inwards or Turning Outwards*, Budapest, Akadémiai kiadó, 1985, pp. 218–32.
6. S. Bulmer, 'The European Council's First Decade: Between Interdependence and Domestic Politics', *Journal of Common Market Studies*, vol. 24, no. 2, 1985, pp. 89–104.
7. J. Kornai, *The Economics of Shortage*, Dordrecht, North Holland Publishing, 1980.

6 The economic development of the CMEA countries and the CMEA itself

The main purpose of this chapter is to give the reader a simple, outline sketch of each of the European CMEA member countries. This is supplemented by a briefer look at the non-European members and a concise account of the development of the Council for Mutual Economic Assistance. Each European member is treated in turn, looking at its growth record, external trade and economic outlook, and concluding with a thumbnail sketch of the evolution of its economic system. The growth of the non-European group is then examined, and some comments on the possibility of further expansion are offered. Finally, statistical tables show the trends in selected basic economic magnitudes.

The European Members

Bulgaria

Although Bulgaria has not been immune to the general trend of declining growth rates, it has been probably the most successful in sustaining quite a respectable pace of growth and at the same time containing its hard currency debt by substantial trade surpluses. It was of course one of the least developed East European countries and encountered the barriers posed by the need to switch from the traditional extensive to a more intensive growth pattern relatively late. State employment was still growing rapidly in the first half of the 1970s, and the transfer from agriculture to industry has also been fast. Investment has generally increased at a high pace too, though erratically in the later 1970s. Most of the time consumption has been accorded low priority, the regime preferring to concentrate resources on high-speed industrialisation. Agriculture has done relatively well and the country's position as a net exporter of food has certainly been an important ingredient in its overall comparative success. The fact that it is a small country and has deliberately developed a distinctly one-sided industrial profile has probably helped to mitigate the disadvantages of administrative central planning until recently.

Bulgaria conducts a higher proportion of its foreign trade with socialist countries than any other European CMEA member, around three-quarters of the total. Over half its trade is with the Soviet Union alone. This is the key to its industrial growth; it has deliberately developed its engineering industries, in particular, to supply the Soviet Union, and of course buys most of the necessary materials from the Soviet Union since the country is poor in natural resources. It is thus very closely integrated with the Soviet Union, and the relatively undemanding quality requirements of its Soviet customers have facilitated its quantitative expansion of output. It has also developed significant exports to the Third World. The remaining feature of Bulgarian trade is its ability to export food surpluses to Western Europe, thus earning valuable hard currency to pay for modern equipment for its industries. In the late 1970s the country was recording substantial surpluses on hard-currency trade and it has no serious debt problem. It should also be noted that although Table 6.6 shows that Bulgarian foreign trade grew less than some other CMEA countries in the 1970s (after the largest increase in the 1960s) these figures are in current prices and prices in intra-CMEA trade rose less fast than world prices in the 1970s. The effect of this is to understate, relatively, the volume growth of the trade of countries which trade most within CMEA, as does Bulgaria. Unfortunately, all CMEA foreign trade statistics are in current prices.

Although the Bulgarian leadership is far from professing itself satisfied with the country's economic performance—criticisms of inefficiency and 'reforms' supposed to improve it have been quite frequent—there is no indication of impending changes in basic policies, either affecting the economic system or the directions of development. The regime apparently expects to continue along the established lines, with continuing efforts to raise efficiency. There are, however, a number of factors which are likely to make the familiar path harder to follow in future.

In addition to the tendency for central planning to encounter greater problems as an economy increases in complexity, the Soviet Union is generally becoming a more demanding customer as far as quality and technical sophistication are concerned. At the same time, Bulgaria's terms of trade *vis-à-vis* the Soviet Union have deteriorated and are unlikely to swing back in the country's favour again. It might even be forced to look to the world market for extra sources of energy or raw materials, though the high degree of integration with the Soviet economy may persuade the Soviet Union to grant Bulgaria preference over some other CMEA countries. Another potential difficulty is that the accession of more Mediterranean countries to the EEC (first Greece and now Spain and Portugal) will increase the competition faced by Bulgaria's agricultural exports to Western Europe, thus making it harder for her to earn the currency to buy modern technology to raise the quality of her industrial products. Bulgaria, therefore, whilst facing less formidable problems than several of her fellow CMEA members, can certainly not afford to be complacent about the future.

Bulgaria has clearly not escaped the familiar weaknesses of the traditional Soviet-type economic system. Complaints of inefficiency, waste, delays in completion of investment projects, low productivity and poor quality are all as common in Bulgaria as elsewhere in CMEA. A very mild flirtation with a profit-orientated reform in the mid-1960s gave way to a return to the traditional central planning and various attempts to improve its efficiency by administrative reorganisation. In the early 1970s horizontal and vertical integration in 'associations' was the preferred policy: in 1976 this was supplanted by emphasis on 'combines', the grouping of enterprises around one central or leading one, often on a locality basis. These moves may have simplified the communication and planning work somewhat, but did not and could not do anything to tackle the fundamental flaws in central planning. Bulgaria remains clearly an administratively planned economy, and it can thus be predicted that as the economy becomes more complex the difficulties of sustaining the rate of growth the regime desires will multiply.

Czechoslovakia

The annual percentage increases in Czechoslovak national income and industrial production show a steadily declining trend throughout the 1970s, culminating in the total stagnation of 1981–2, the worst figures since the actual fall in national income in the early 1960s which precipitated the reform movement of that period. Agriculture maintained a distinctly creditable rate of growth up to 1977, benefiting from reforms made before August 1968 and not subsequently rescinded— abolition of or major reductions in compulsory procurements, increased decision-making autonomy for farms, and larger increases in the prices paid for agricultural produce than in those charged for industrial inputs bought by farms, all of which improved pay and incentives in the sector. This agricultural success had relieved the balance of payments considerably, but the poorer performance in 1978–81 placed a new burden on the economy as livestock feed had to be imported from the world market; more recently there has been some recovery from these lean years.

The reason for the falling rate of growth of the Czechoslovak economy is essentially the failure to make the transition from Soviet-type extensive growth, based on increasing use of all inputs, to growth based on increasing efficiency and productivity. The trend in labour-productivity growth was in fact downward during the 1970s, dropping from an average rise of 5.1 per cent per annum in 1966–70 to 3.0 per cent per annum in 1976–9. The rate of increase of the labour force has been steadily dropping and the growth of the population is slow, so that there is little scope now for further increases in employment, as the agricultural sector is quite small and the participation rate high. Investment too rose only slowly in the second half of the 1970s and fell in the early 1980s. The country thus exemplifies all the elements in the general crisis of the Soviet-type growth strategy. All the familiar problems are to be found: dispersion of investment, unresponsiveness of production to demand, excessive use of materials and labour, despite the scarcity of both. In addition, Czechoslovakia has a particular problem deriving from its position as one of the most advanced producers of capital goods in the CMEA; the other countries' demand for Czechoslovak engineering products has led to a high degree of concentration on this sector, although the country lacks the raw materials of which the industry is a heavy user. Now, too, it increasingly needs to import sophisticated equipment and technology from the West in order to modernise its product range to meet its customers' requirements.

The proportion of Czechoslovakia's foreign trade that is with socialist countries is a little lower than Bulgaria's, but it is still very high at 70 per cent, half of which is with the Soviet Union. During the 1970s the proportion of imports coming from the developed West was higher than that of exports to these countries, and there were steady, though not massive, deficits on hard-currency trade in the second half of the 1970s. After 1980, Western imports were cut back in order to contain the country's hard-currency debt to a moderate level. Czechoslovakia's problem is that it has an industrial structure which is material- and energy-intensive, but it has few natural resources (some coal, much of it brown, and also uranium). Its terms of trade therefore worsened during the 1970s. It is now uncertain how far it will be able to rely on its traditional supplier, the Soviet Union, for increasing energy requirements, for instance. Yet if it were to try to change its industrial structure it would find this difficult because of the other CMEA countries' demand for Czech engineering products. So the country is trapped in the CMEA system of integration through bilateral trade agreements between governments.

The absence of a serious debt problem is virtually the only bright feature on the otherwise gloomy economic horizon of Czechoslovakia. Domestically, there is nothing to suggest that the trend of falling growth rates will be reversed, as the reasons for the trend have been clear for some time, and the efforts made hitherto to improve the situation have not worked, but there is no indication of willingness to tackle the fundamental malady by a thorough reform of the economic system. The pressure of investment by itself is one important factor depressing Czechoslovak living standards, and this competition for resources is exacerbated by the country's external trade position, in which it has to export more manufactures to pay for its raw materials. The key to the future of Czechoslovakia is probably Soviet ability or willingness to supply increasing quantities of energy and raw materials. As long as the Soviet Union does so, albeit at rising prices, Czechoslovakia is likely to continue on the same basic path, with stagnant living standards. If, however, the Soviet Union were to insist that Czechoslovakia turn to the world market for additional energy, in particular, this would create great pressure for change in its industrial structure and also for reform of the economic system so as to make Czechoslovak industry more competitive in world markets.

Before the Second World War Czechoslovakia had already reached a level of industrialisation comparable with Austria at the same period. Many Czechoslovak economists consider that in view of this fact the adoption, or imposition, of the traditional Soviet-type planning system at the end of the 1940s was a fundamental mistake. It was probably because of this particular inappropriateness of the system that Czechoslovakia reached an economic crisis, with negative growth, by the early 1960s. Yet because the ensuing moves towards thorough economic reform were accompanied by a political liberalisation which was unacceptable to the Soviet Union, the 'normalisation' after August 1968 included a return to the administrative planning system. The 1970s saw various attempts, by reorganisation and revision of indicators and regulations, to make the system work more effectively, but although there was some increase in the use of financial instruments the trend was to greater centralisation. In 1978, it is true, the 'complex experiment' was introduced in some 100 enterprises, devolving a larger measure of autonomy to their management within the framework of supposedly fixed three- or five-year plans. Initially it did not seem that this was to become general but in 1980 a resolution 'on the improvement of the system of planned management' was adopted. However, while this did represent a move towards decentralisation, the principle remains the distribution of centrally prescribed indicators to enterprises. In essence, therefore, the planning model is unchanged and it seems in the highest degree unlikely that the present Czechoslovak leaders would venture on any reform reminiscent of that proposed in the mid-1960s.

East Germany

Generally speaking, during the 1970s the record and position of East Germany were similar to those of Czechoslovakia, though the performance of agriculture was subject to greater annual fluctuations in the former; but this sector is of relatively minor importance in both countries. The trend of growth in East German national income, industrial output and investment was unmistakably downward, though, unlike Czechoslovakia, this decline was successfully halted from 1980; investment growth, however, has been very slow since then. In most years the increase in employment was the lowest for any CMEA country. The population actually declined slightly in most years in the 1970s and even in 1960 the rate of increase was so low as to ensure an almost static labour force from now on. East Germany is thus the most extreme case of the labour-shortage barrier to growth on the 'extensive' pattern. The country is also most conscious of pressure stemming from the gap between West and East European living standards because of the population's greater awareness of the situation in West Germany owing to contacts with relatives and the ability to receive and understand broadcasts.

East Germany, like Czechoslovakia, is suffering from the inflexibility of administrative planning in a complex modern economy; it is the most industrialised CMEA country and it, too, lacks raw materials but has nevertheless concentrated on heavy industry. The reason again is that because of its initially higher level of development it was able to supply equipment greatly needed by the less advanced CMEA members, above all the Soviet Union.

Until 1980, East Germany's foreign trade pattern was also similar to that of Czechoslovakia, with almost the same percentages of turnover with socialist countries and the Soviet Union. But one significant difference was that, whereas for Czechoslovakia the geographical distribution of exports and imports was not very dissimilar, for East Germany the proportion of exports to socialist countries was substantially higher than the proportion of imports coming from those countries (though this does not apply to trade with the Soviet Union; the figures may reflect East German deliveries to non-European socialist countries). At the same time, the percentage of GDR imports coming from the developed West was much greater than that of exports to that area. This was reflected in the large hard-currency trade deficits registered by East Germany in the second half of the 1970s, and the substantial level of debt accumulated.

Since 1980 there has been a notable drop in the proportion of exports going to socialist countries (while in Czechoslovakia it has increased) and a corresponding rise in the proportion going to the developed West. This has been a necessary and deliberate policy move in order to tackle the Western debt which has been incurred. The geographical distribution is now very similar for exports and imports. Of course, the country's special trade relationship with West Germany makes it unique in the CMEA, but it cannot be assumed that this source will provide unlimited credit, and it is also a cause of additional pressure on the regime.

It is hard to see any prospect of relief from the severe pressures besetting the economy. There is no indication of readiness to tackle the weaknesses of the economic system seriously, although their consequences are clear. The Soviet attitude concerning energy and raw material supplies is crucial, as the country would experience even greater difficulty if it also had to finance rising imports of these items for hard currency. The static labour force means there is great pressure for more capital investment to raise productivity, but the latter also depends on incentives for the labour force which constantly compares its standard of living with West Germany's much higher level and feels dissatisfied.

The New Economic System in operation in East Germany in the mid-1960s involved a significant measure of decentralisation, and enterprises responded by using their new autonomy to pursue lines of production (and investment, to the limited extent that this was in their power) which offered the most profit. Unfortunately the planners were unable to reform the price system in such a way that what was profitable for enterprises was also what the regime wanted to see produced. The outcome at the end of the decade was recentralisation by means of 'structure-determining tasks' which enterprises had to fulfil. During the course of the 1970s there was some administrative change, with a decline in 'associations', an intermediate administrative body between industrial ministries and enterprises, and a corresponding growth of 'combines', which are now nearly universal in East German industry. This change represents in one way a step towards centralisation, because the combine has much more power over its constituent plants than associations did over their enterprises. In another way it represents, or was intended to represent, a decentralising move, with some of the powers of ministries devolved to combine level. But the system remains fundamentally one of administrative planning, and in particular there has been no effective price reform, without which the combines could not be given extensive autonomy. One of the crucial reasons for the failure to tackle the prices problem is that it would have to include increases in many retail prices, and the regime is evidently nervous about such a move. Discussions continue in East Germany on how to improve the efficiency of the economy, but there is no sign of readiness for fundamental change on the part of the leadership.

Hungary

From a glance at the basic statistics the position of Hungary seems in no way exceptional; declining growth of national income and industrial output, marked fluctuations in the rate of increase of agricultural production and of investment. Furthermore, the latter fell consistently in the early 1980s. Population growth has been very low for the past two decades, exceeding only that of East Germany, so labour is in short supply too. Nevertheless, there is little doubt that the Hungarian consumer is better off and more satisfied than is the case in any other CMEA country. A consistent element in Hungarian economic policy ever since the aftermath of the 1956 revolution appears to have been to give rather greater priority to consumption than has been usual in other CMEA countries, though even this cannot be shown by comparing the aggregate proportions of national income invested and consumed during the 1970s. The explanation seems to be the greater flexibility and responsiveness of the Hungarian economy to consumer demand in details. The 'unregistered' economy is also thought to be most extensive in Hungary, as a result of the relative decentralisation; this of its nature cannot be shown statistically, but it undoubtedly tends to work to the benefit of the consumer. For intermediate goods too, the replacement of output plans and centralised allocation by enterprise choice of product mix and distribution through wholesale trade has brought much-needed flexibility and responsiveness to customers' real requirements, despite the near-monopoly position of many producers. All the same, the economy is now growing slowly and facing many of the same difficulties with raw materials and terms of trade as its neighbours, and during the first half of the 1980s real incomes have been declining as retail prices rose faster than wages.

Another point, which applies to all the statistics for economic growth rates in the CMEA countries but is particularly relevant at this juncture, is that they are widely believed to overstate the true rates of growth, but by differing margins in different countries. A United Nations study[1] which compared the official growth rates with those obtained by applying the real product methodology in both Western market economies and East European centrally planned economies found that in the former the difference between the two sets of figures was usually about 0.5 of a percentage point either way, whereas in the latter the growth rates obtained in the study were consistently lower than the official ones. For Czechoslovakia and Bulgaria the discrepancy was around 1 percentage point, for Poland and the Soviet Union 2 percentage points, and for Romania and East Germany as much as 3 percentage points. For Hungary, the official growth exceeded the study's figure by only 0.6 of a percentage point, not substantially more than in the case of the market economies of Western Europe. This is very significant for comparisons between Hungary and other CMEA countries, in particular East Germany, helping to explain why the universal subjective impression of better consumer-goods supply and standard of living in Hungary does not show up in the comparison of official growth statistics.

The three countries so far examined, Bulgaria, Czechoslovakia and East Germany, all showed a rather similar geographical distribution of their foreign trade, though the consistent concentration on the socialist countries was greatest in the case of Bulgaria. The Hungarian trading pattern is noticeably different, and the difference increased markedly during the second half of the 1970s. A considerably greater share of trade is now with the developed West, and there has been a significant rise in the proportion with LDCs too. This trend reflects Hungary's efforts to align its economy with the world market in addition to the general CMEA policy during the 1970s of importing modern technology. It also reflects the difficulty of reconciling Hungary's socialist market economy with the central planning, materials allocation and bilteral trade agreements of the other CMEA members; this problem is discussed in Chapter 8.

Hungary had a consistent though not rising deficit on its hard-currency trade and its total debt increased gradually during the 1970s, but the debt-service ratio remained manageable. The country has had to exercise strict restraint to keep the situation under control and this has resulted in a fall in the proportion of imports from the developed West in the early 1980s. Hungary suffered distinct deterioration in her terms of trade during the 1970s and the higher cost of energy and raw materials was one of the principle causes of the slow growth from the second half of the decade. In 1982 the country suffered a temporary crisis in respect of borrowing from the West as a side effect of the Polish situation. There was a failure on the part of Western bankers to appreciate the superior quality of Hungarian national management compared with that which had produced the Polish debt disasters. This was eventually resolved, and whilst Hungary's debt is a significant restraining factor in its economic policy, it is not generally viewed as a problem country in this respect.

Because of its poor natural endowment and dependence hitherto on Soviet supplies, Hungary too must be concerned about the extent and the terms on which this source will meet future requirements. The policy of orientating trade, and ultimately the development of the economy, more towards the world market will, however, mean that the country is better prepared if it does have to turn to non-CMEA suppliers. Hungary is also fortunate in being able to produce enough food both to supply the home market adequately and to export meat to the Soviet Union, for which at present it receives hard currency. Nevertheless, it looks unlikely to be able to sustain a rate of growth faster than the average of the latter half of the 1970s, owing to the combination of a tight supply situation for both raw materials and labour and the worsening in its terms of trade. Since it is essential that investment in raising industrial efficiency and the technological and quality standards of Hungarian manufactures should proceed if the country is to fulfil its declared aim of matching world-market requirements, the scope for increases in consumption is severely limited.

Hungary is the one CMEA country which carried out a complete reform of the traditional system of central planning. The New Economic Mechanism brought into operation in 1968 represented a fundamental break and its essential features, the abolition of compulsory plans for individual enterprises and of the materials allocation system, have not been changed subsequently. The third crucial element in the NEM, the movement towards a market-type price system, while never intended to be anything but gradual, has advanced more slowly than the designers of the reform hoped. It was also an area in which some backtracking occurred during the 1970s, when there were periodic bouts of partial recentralisation, including tighter regulation of enterprises' wage payments and investment expenditure and continued retention and even extension of the numerous special price subsidies and taxes. These are called 'financial bridges' by the Hungarians, because their function is to bridge gaps between existing Hungarian and world market prices, and the aim of the NEM was gradually to eliminate them.

A renewed drive to do this commenced at the start of 1980 with a new price reform. Its aim, in the words of the President of the Materials and Prices Office, was a 'transition from the prime cost price system to a competitive price system', the latter being one where 'the input price of natural resources is determined by the import price and the final product price is determined by the export price in non-rouble terms'. Large increases in the retail prices of food and energy were involved in this adjustment process, which is not yet complete; income supplements were given to compensate for the effect on the cost of living. The price reforms were a clear indication of Hungarian determination to continue on the path of a socialist market economy and, in particular, to take the criterion of profitability at world-market prices as the basic guide to investment and thus ultimately to the structure of the Hungarian economy. It is hoped that the economy can then be exposed to foreign (Western) competition to check abuse of monopoly power by domestic producers. The Hungarian reform is discussed more fully in Chapter 8.

Poland

Poland's distinctiveness is clear from the basic statistics (Tables 6.1, 6.2, and 6.3): extremely rapid growth in the first half of the 1970s was followed by collapse by the end of the decade in both industry and agriculture. The increases in investment in the years 1972–5 were even larger, and were followed by a similar fall. Much of that investment was of course financed by foreign borrowing to buy imported equipment. The labour-supply situation in Poland is also different from that in the countries considered hitherto. The population has continued to grow at a rate fluctuating slightly around an average of about 1 per cent per annum and there has been concern about the difficulty of providing enough jobs for new entrants to the labour force. Employment grew rapidly in the early 1970s, but towards the end of the decade the rate slowed greatly owing to declining job opportunities, rather than labour scarcity, and in 1981–2 employment actually fell.

The disruptions in Poland in 1980–1 are of course well known, and were one cause of the economic collapse, but this had started earlier. One reason was that many of the new plants created with the equipment imported in 1972–5 also needed continuing imports of materials and components, and as the rapidly growing debt forced cuts in imports many such plants could only operate far below capacity. It is also clear that there was a gradual but widespread breakdown of the system of planning and co-ordinating the economy, which had in any case become seriously overheated as a result of the preceding investment boom. Inflation became more pronounced and relatively open, for a centrally planned economy, because of enterprises' powers, in a chronic sellers' market, to set their own prices for new products. The stagnation of agricultural output led the government to try three times, in 1970, 1976 and 1980, to raise retail prices to reduce the excess demand, with the now familiar political consequences. The ambitious designs of the early 1970s were probably always unrealistic but the changes in the world economy after 1973 completely scuppered any chances of success which they might otherwise have had by making it far more difficult for Poland to break into world markets for manufactures in the way she needed to to repay her borrowings.

Total Polish foreign trade turnover grew very rapidly in the years 1972–5, reflecting the massive imports of capital goods on credit. Thereafter the overall growth was much slower, but the hard-currency deficits continued to be large, although their trend was declining after 1976. The total debt and the debt–service ratio, however, went on rising, resulting in the well-known payments crisis and rescheduling negotiations. Poland, like Hungary, already conducted a lower proportion of its trade with socialist countries in 1970 than did Czechoslovakia and East Germany, let alone Bulgaria, and the proportion was at the same level as for Hungary in 1980, too. But the trend during the 1970s was different; whereas during the first half of the decade Hungarian trade swung back somewhat to the socialist countries, and then sharply to the West, Polish trade was most West-orientated in 1975. The proportions diverged widely for imports and exports; almost half of the former coming from the developed West in 1975; by 1980 the Western share was just over one-third in both directions. By 1983, however, Polish trade had swung back much more to the socialist countries, and the change, whilst large overall, was most marked in imports, because of the debt-enforced cutback on purchases from the West.

Poland's terms-of-trade experience had been different from that of the other five small CMEA countries of Eastern Europe. She is an important coal producer and exporter, and used also to export meat products until the pressure of home demand on stagnating production curtailed this trade. These advantages offset the losses caused by rising prices of other forms of energy and raw materials which Poland has to import, and the terms of trade fluctuated somewhat but on balance have not altered significantly. Poland's problem in this respect is to produce enough of these traditional export items; this was the reason for the prolonged dispute above the five-day week in coal mining and the military regime's reintroduction of six-day working. The development strategy of the early 1970s relied on gradual modification of the commodity structure of exports, paying for the imported plants with manufactures produced by them. Now many of these plants are either still unfinished or working far below capacity because of shortages of (often imported) inputs, and the export prospects for their products are in any case poor. There are repercussions for other CMEA countries as well, since some of them have relied on Polish coal and they constitute another source of pressure to raise output; domestic demand, too, is not fully met, with consequent shortages of electric power. Food shortages, in the sense of queues and unsatisfied demand, should have been reduced greatly by the extremely large retail price increases put into effect at the end of January 1982, but wages also continued to rise very rapidly, partly negating the effect.

Poland's economic policy is going to be dominated by the foreign trade constraint for many years. Imports will have to be tightly controlled and great efforts made to increase both the range of products which can be sold for hard currency and the level of production of these items. Living standards have fallen sharply and cannot be expected to recover to the level of the mid-1970s for a considerable time. The growth of investment resumed in 1983, after falling by over 50 per cent in 1979–82, as efforts were made to complete unfinished projects and to raise coal and power output. Large-scale imports of investment goods must be at an end for the time being.

As far as reform of the economic system is concerned, Poland is something of a paradox. It was the first CMEA country to start the process, as early as the autumn of 1956; it has a tradition of producing first-class economists (Oskar Lange and Michal Kalecki, for example) which the more recent generation has maintained; there has been continued discussion of the need for reform and several attempts to introduce reform packages. The New Economic and Financial System of 1973 was undoubtedly influenced by the Hungarian NEM but failed to heed the crucial lessons of the Hungarian experience, namely the importance of comprehensive introduction of the new system throughout the economy at the same time and the need for appropriate preparation of the economic environment. The system blueprint itself was also less consistent and showed signs of greater compromise, for instance in adopting a dual rather than a single criterion of enterprise success (net profits *and* net value added, in contrast to the Hungarian use of profit alone). The outcome was that because the reform was brought in during the period of boom, with universal excess demand, and because the traditional system was still operating alongside it, it soon ran into difficulties, which provoked efforts to reassert greater central control. These, however, were not very successful, partly because enterprises retained some power of autonomous action, and partly because of the general political degeneration of the regime in the late 1970s. By the close of the decade Poland had neither a reformed nor a functioning traditional central planning system; the economy had become a juggernaut out of control.

Soon after the introduction of military rule, Jaruzelski spoke several times of the intention to proceed with reform. Officially, a reformed system, providing substantial enterprise autonomy along lines similar to the Hungarian model, has been introduced. Yet it is very difficult to put such a reform into effect in the tightly constrained economic environment of Poland in the early 1980s. The concentration on investment and debt repayment at the expense of consumption which for the time being is essential for the country, is likely to result in continued intervention from the centre. This is discussed further in Chapter 8.

Romania

During most of the 1970s, Romania sustained high rates of national income growth—in many years the highest of all the East European countries. This is particularly noticeable in the second half of the decade, when growth declined in the other countries before this trend showed itself in Romania; by 1980, however, the country was displaying the common pattern. The industrial sector generally mirrors the overall trend of the economy; agriculture, on the other hand, was most notable for its extreme fluctuations, with three years of very big increases in output (1971, 1972 and 1976), and virtual stagnation if not decline in the remaining years. Growth has been very much the priority of economic policy, with concentration on investment, which increased rapidly up to 1979. The natural increase of the population was falling somewhat by the end of the decade but prior to then it had been around 1 per cent per annum, and the growth of state employment was rapid compared with all the other countries except Bulgaria. This is not surprising as these two countries certainly were the least developed in Eastern Europe, and Romania still had 29 per cent of its labour force in agriculture in 1983. Romanian growth has thus been of the 'extensive' type, with rapidly rising inputs of labour and capital. Consumption has been neglected and living standards are low; this has caused occasional outbreaks of worker protest. While these have been nowhere near the scale of those in Poland, the government must be conscious of growing pressure for improvement in the level of consumption.

Romanian foreign trade turnover grew appreciably faster than that of any other CMEA country during the 1970s; the figures, however, are in currrent prices and because the proportion of Romanian trade which is with CMEA countries is the lowest of all, the element of price inflation was probably greater in the Romanian case. (This is because, in principle, intra-CMEA trade prices follow world market prices with a lag because the general basis is a five-year moving average.) After 1980 the growth of Romanian trade came to a halt, with a big drop in 1982. Romania had moved by the end of the 1960s to a geographical spread of trade close to that reached by Hungary and Poland a decade later, and the trend away from the socialist countries continued during the 1970s, with the fastest growth taking place in trade between Romania and LDCs, which by 1980 accounted for a much larger proportion of Romanian than of any other European CMEA country's trade. The relative weight of trade with LDCs was greater in imports than in exports in 1980, though not earlier in the decade; this reflects Romania's policy of moving away from dependence on Soviet raw materials and oil after the disagreements of the early 1960s concerning development strategy and Romania's role in the CMEA pattern of integration. (Romania is an oil producer, but domestic production is no longer sufficient for requirements and is supplemented by imports from the world market.) The proportion of Romanian trade with the Soviet Union is conspicuously low. After 1980 there was a small shift back to the socialist countries in the distribution of Romanian exports, and a much bigger one in the case of imports. This reflects cuts in Western imports because of currency shortages and the need to service substantial hard-currency debts built up during the late 1970s, especially 1979 and 1980, when there were large deficits on trade with the West.

The reason for the change in the direction of trade during the late 1960s and the 1970s was evidently at root political; it was both an economic form of the general Romanian declaration of independence and a precaution to reduce the country's vulnerability to economic pressure from the Soviet Union; it may also have been caused by actual Soviet refusal to supply some raw materials. In relation to its political objectives the policy has been successful in that by 1980 less than one-fifth of Romanian trade was with the Soviet Union. It has, however, involved the cost of direct exposure to world price movements without the cushioning delay provided by the CMEA pricing mechanism.

Romania started as one of the least developed countries in CMEA and, where its domestic economy is concerned, it may be described as less far along the road to the same problems which face the more advanced member countries. Labour transfer from agriculture still holds considerable potential for 'extensive growth', though it will be harder to maintain rapidly rising investment in view of the twin pressures of increasing discontent with poor living standards and the balance-of-payments constraint. If world energy and raw materials prices were to rise steeply again, Romania would be hit hard, though for the next few years this does not look very likely. The growth slowdown may not be as severe in Romania as in most CMEA countries yet, but the trend is unmistakable.

Romania showed the least interest of all CMEA countries in economic reform in the 1960s and only belatedly announced a so-called reform which brought about little or no change in the centrally planned system. During the 1970s there were administrative reshuffles and another so-called New Economic Mechanism in 1978 which once again was quite clearly intended only as an effort to make central control more effective. In its economic policy the Ceauşescu regime has been totally orthodox and adhered firmly to the Soviet model of central planning introduced into the country more than three decades ago. The high degree of personalisation of the political regime under Ceauşescu seems to have extended to the sphere of economic administration, too; a study of Romanian industrial enterprises in the early 1970s by Granick concluded that managers were rewarded not so much on the basis of formal plan fulfilment as of personal assessment and recognition by their ministerial superiors; when necessary, plan targets were adjusted to fit these assessments. The relative economic 'backwardness' of the country, combined with the distinctive political regime, mean that the prospect of systematic change in the economy looks remote.

The Soviet Union

Although the Soviet Union differs greatly from the six smaller countries in having a rich endowment of natural resources, its growth record is very similar to that of the other countries. The rates of increase of national income and industrial output declined steadily throughout the 1970s, though there has been a slight revival since 1982. Agriculture has performed erratically but on average very poorly, with the growth of output barely equalling the increase in population during the 1970s. Investment growth almost ceased at the end of the 1970s, and these figures may overstate the true growth of investment because of rising costs not reflected in the Soviet price indexes (this is discussed in more detail in Chapter 7). The rate of increase of the state-employed labour force has fallen steadily and labour is now in short supply in many areas, though not in Central Asia. But the people there are unwilling to move elsewhere, and it is the high birth rate in the Moslem areas of the country which account for most of the overall population increase; the Slav (and Baltic) population is almost static, so the labour-supply position is much worse than the overall figures indicate. The country has been striving with little success to achieve the much-discussed transition from extensive to intensive growth, as it is no longer in a position to continue pouring in more and more labour and capital; it is now essential to raise labour productivity. A major obstacle to this is the inadequate supply of consumer goods, both food and non-food; the population has accumulated large cash reserves and savings-bank deposits because more has consistently been paid out in wages than can be spent in the shops. The promise of paper-money pay is therefore no incentive now and will only become an effective one if there are sufficient goods for sale. This is a difficult condition to fulfil because of the intractable problems of grain production, compounded by very inefficient use of feed grain in the livestock sector, and the inflexibility of the economic system, which has an endemic propensity to concentrate resources on heavy industry. The government's concern with military power is obviously also a vital factor which precludes a substantial switch of emphasis to consumption.

Soviet foreign trade turnover in current prices grew faster between 1970 and 1983 than that of any other CMEA country: the years of greatest increase were 1973–5 and 1979 onwards, reflecting the big rise in world prices of the energy products and raw materials of which Soviet exports predominantly consist. These financed imports of modern technology and, in several years, of large quantities of grain.

The trends in the regional distribution of Soviet trade are interesting, though allowance should be made for the more rapid rise in world than CMEA trade prices during the 1970s. The share of the CMEA countries declined quite markedly, with the reduction occurring in imports before it did in exports; this reflects the delayed rise in CMEA energy prices. Soviet trade with socialist countries, however, accounted for a larger proportion of total trade in 1980 than it did five or ten years earlier. There are two factors at work here: Soviet trade with Vietnam, for example, rose greatly and, in addition, some countries which were formerly classified as LDCs had been reclassified as socialist countries by 1980. The proportion of Soviet trade with the developed West rose steadily, whilst that with LDCs appeared constant, but rose somewhat when allowance is made for the changes of category.

The Soviet debt position has been kept well under control, though this is not to say that payments can be balanced without holding imports below the levels that might be desired by the Soviet planners. The recurrent need to buy large amounts of grain for hard currency following bad harvests must displace other imports, mostly modern equipment. The weakness of the world market for oil, gas and raw materials must be cutting earnings from these Soviet export staples. The fall in the gold price also adds to the Soviet Union's difficulties. It is true that fiercer competition between Western suppliers in the current world market might be expected to lower prices of Soviet imports, too, but not to the same extent, and in any case the Soviet Union is probably not in a position to take advantage of such possibilities because of the prior commitment of funds to grain purchases.

The Soviet economy faces the full range of constraints on future growth familiar in this survey of CMEA countries. In most parts of the country, particularly those where industry is concentrated, labour is now in short supply. Investment funds are under great pressure, too. There is quiescent if not open consumer dissatisfaction and a lack of interest in financial incentives as long as they are not backed up by a great improvement in the supply of consumer goods in the shops. A monetary reform of some kind, to wipe out much of the accumulated purchasing power, is regarded as a possibility by some observers. In these circumstances economic growth is unlikely to accelerate very much from the current historically low level, though it is probably wrong to predict a further decline into total stagnation and even contraction simply on the basis of extrapolation of the trend of the past.

The Soviet Union's rich deposits of natural resources are obviously a source of economic strength in the long run. But because of their location in underdeveloped, remote and inhospitable terrain they demand heavy investment before production can be increased. In many instances the levels of output now sustainable are insufficient to cover growing domestic and CMEA demand and provide enough for export to hard-currency markets on a large scale. If production capacity is not expanded now, the country will not be ready to take advantage of any future rise in world demand and prices. These, of course, are the reasons for joint CMEA financing of some projects and for East–West deals involving sale of equipment against future payment in product.

Agriculture is a crucial problem: as long as levels of output cannot be raised (and costs reduced) there will be the recurrent burden of large grain imports and it will also be impossible to achieve higher living standards and levels of consumer satisfaction, so important for incentives to higher productivity. This problem has to be solved domestically; imports cannot provide more than relief from the worst effects of bad harvest years; it is unrealistic to think in terms of a permanent large foreign contribution to Soviet food consumption. World surpluses are not really large enough in relation to Soviet demand, and the Soviet Union could not pay for them even if they were. Agriculture's problems are partly peculiar to the sector and stem ultimately from the forced collectivisation and the years of neglect and deprivation which followed it. They are also the problems of the whole economy—inefficient use of resources, unresponsiveness of producers to customers' requirements, and wasteful inflexibility.

The Soviet Union introduced a cautious economic reform in 1965, which was to be extended gradually and never envisaged more than a strictly limited degree of devolution of decision-making to enterprises. In fact a combination of inconsistencies in the reform itself and opposition to any dilution of their erstwhile powers by the economic administrators resulted in frustration of even these modest intentions. The reform was not abandoned; on the contrary, the 'new system' became general. But the system made no noticeable difference to the manner in which the economy functioned; it remained a centralised, administratively planned economy.

This was followed by repeated reformulations of the managerial incentive system, which gradually became almost incomprehensible and consequently could not be effective. There was also an administrative reorganisation in 1973, when most enterprises were merged into associations. This, like the corresponding measures in many other East European countries, was designed to simplify the structure of economic administration but did not affect the principles of central planning.

The reform in 1979 was once more inconsistent, complicated, and did nothing to touch the fundamental defects arising from the issue of compulsory targets to enterprises and the accompanying system of rationing materials. The principle of central setting of prices was unaltered. Similar criticisms apply to the 'experiment' initiated in 1984 and extended in 1985, of which more in Chapter 8. The decade and a half since 1970 witnessed successive tinkering with the central planning system, but no suggestion of recognition that it has fundamental weaknesses which cannot be eliminated in this kind of way. There has been evident dissatisfaction at the highest level with the performance of the economy, but, in Brezhnev's later years at any rate, only a narrow, blinkered view of the way to tackle the problems. Andropov had some success in his campaign for better work discipline and against corruption, but was an effective leader for less than a year. Gorbachev is clearly much more energetic, has much more time, and may achieve relatively better results, but the signs are that he is not considering fundamental systemic change but simply a more determined drive to improve the efficiency of the existing system.

The Non-European Members

Mongolia

Initially the constitution of the CMEA confined its membership to European countries. This restriction was eliminated in 1962, when the Basic Principles of the Socialist International Division of Labour were adopted and Mongolia became the first non-European member.[2] This was the time when Khrushchev was seeking to introduce supranational planning as the way to integration of the member countries' economies, and encountering resistance to the loss of sovereignty this entailed in Eastern Europe, most vehemently from Romania. The Basic Principles contained an explicit commitment to the levelling-up of the poorer countries with the help of aid from the richer ones. Mongolian trade had long been and has continued to be heavily concentrated on the Soviet Union, though trade with China became substantial during the 1950s, so that by 1957 'only' 75 per cent was with the Soviet Union. As

Soviet–Chinese and thence Mongolian–Chinese relations deteriorated, the proportion of trade with China dropped again, so that by the mid-1960s the almost exclusive concentration on the Soviet Union was restored. Aid, mainly Soviet, in the form of long-term low-interest credit financed around half of the country's investment during the late 1960s and the 1970s. This is an extremely high proportion by any standards. The essence of the Mongolian relationship with the CMEA, and within it principally the Soviet Union, was and is readiness to accept integration by coordination of planning, with the concomitant loss of sovereignty, in return for substantial aid, in the expectation of thereby accelerating the country's rate of development and approach to the income level of the European CMEA members. It is thus an apparently enthusiastic participant in the Agreed Plans, and target programmes where they involve it.

It must be remembered that, despite its large territory, Mongolia has a tiny population—less than 2 million—although the growth rate is rapid (see Table 6.5). It also has large numbers of livestock, as well as much land, per head, so that even in 1962 it was not a particularly poor country by the standards of the Third World. The cost of the commitment to bring it up to the income levels of the European CMEA members was thus not excessive. In this respect the country was perhaps comparable to one of the smaller Central Asian constituent republics of the Soviet Union, or a Siberian province. The fact that it has a common frontier with the Soviet Union is very significant. The timing of its consolidation into the Soviet bloc which joining the CMEA meant was also undoubtedly influenced by the Sino–Soviet split; from 1962 Mongolia was clearly aligned with the Soviet Union in that dispute.

Cuba

The next non-European country to join the CMEA was Cuba, although it did not do so until 1972, despite having been a socialist country and received Soviet economic support for over a decade—and despite Khrushchev's notorious missiles adventure there in 1962. Cuba is not a big country—its population is now approximately 10 million—and it was not exceptionally poor either. Nevertheless, the cost of the commitment to income levelling implied in the new Basic Principles in 1962 would have been greater than in the case of Mongolia. There are at least three other factors which were probably important in causing the Soviet Union to move cautiously in respect of possible Cuban membership of the CMEA. The first is simply the geographical remoteness of the country from the Soviet Union; all the other members, including Mongolia, formed a single land block (and all except Bulgaria and East Germany share a common frontier with the USSR). The inclusion of a physically distant country can be seen as involving a change in the nature of the group which may well have prompted lengthy consideration. Second, although socialist, Cuba in the 1960s was in various ways distinctly unorthodox, at least when viewed from Moscow. Castro's ideas, and particularly Guevara's, were by no means purely Soviet Marxist–Leninist, and it was unclear how reliable the country would be. Third, there may have been a psychological inclination to caution over Cuba as a consequence of the humiliation of the missiles incident in 1962.

As far as the economic aspect is concerned, the most notable feature of Cuban relations with the CMEA is the purchasing of its sugar at what are generally believed to be well above world market prices. Cuba also supplies nickel, one of the few metals with which the Soviet Union is not particularly richly endowed, and of course buys Soviet oil at below world prices (though the advantage conferred by the CMEA pricing formula has diminished in the most recent years). Cuba is now a participant in the Agreed Plans and target programmes and thus in the general procedure of integration by plan co-ordination.

It is generally reckoned in the West that the cost of supporting the Cuban economy with aid and subsidised supplies and sugar purchases is very high. The benefits for the Soviet Union are political: Cuba has been astonishingly successful in presenting itself to the Third World as 'non-aligned', although it is obviously not, and this has given the Soviet Union a valuable instrument in international diplomacy in the Third World. Cuba has also provided troops to act on behalf of the Soviet Union in Africa—particularly in Angola—where it would have been politically more difficult—and domestically unpopular—to have used Soviet or other CMEA countries' troops alone (although 'advisers' are in Angola and taking an active role).

The timing of Cuba's accesssion to the CMEA is not so clearly linked with specific political circumstances as was the case with Mongolia (or was to be again with Vietnam), though it did fit into the general pattern of a more widely-ranging expansive Soviet foreign policy in the 1970s following Brezhnev's consolidation of his hold on power and the military build-up which began near the end of the 1960s. A further factor may have been the wish to pass some of the cost of supporting Cuba onto the shoulders of the other European CMEA countries.

While Cuba was not in the organisation, its formal relationship was principally with the Soviet Union; once it was a full member there was greater pressure for, and less basis to resist, participation by other members in the various forms of aid to the country.

Vietnam

The last non-European country to become a full member of the CMEA was Vietnam, in 1978. It seems clear that this was a decision reached quickly, in response to developments in Vietnamese relations with the Soviet Union and China, and the hostility of these two, following the end of the war in Vietnam and the takeover of the southern part of the country by the communist regime in the North. The country had joined the CMEA banks, IBEC and IIB, in 1977, but was also a member of the IMF and the World Bank. The 1978 move to full CMEA membership meant alignment with the Soviet Union against China, although Vietnam has continued its non-socialist links with Soviet approval. For the Soviet Union it was particularly important to secure an ally in South-East Asia, close to China; for Vietnam the Soviet Union and the CMEA looked the most promising source of economic aid.

From this latter point of view the reception of Vietnam into the CMEA fold was a very much more serious matter than either Mongolia or Cuba, though by the 1970s the earlier emphasis on the levelling-up commitment had been greatly muted anyway. In terms of population Vietnam is a large country, the second largest in the CMEA, with approximately 60 million people. It is also much poorer than the other two non-European members were at the time of their accession. The cost of a worthwhile volume of aid is thus much more burdensome for the European members. This is doubtless why the country continued to seek aid from Western sources as well, and is encouraged to do so by the Soviet Union. It is also why, according to reports which cannot be definitely confirmed, Czechoslovakia, at least, was opposed to the admission of Vietnam. The country's main exports are tropical crops, which have not traditionally been high on the list of European CMEA countries' import priorities, so the economic benefits for the existing members were hard to identify and the potential costs substantial.

Mozambique's application is refused

Mozambique is a country which would have liked to become a member of the CMEA but is not one. In 1981 it applied for membership, was refused, reapplied and was evidently refused again. This was one of the reasons that led it to sign the Nkomati agreement with South Africa in 1984, under which trade relations were to be normalised. Mozambique undertook to cease supporting the ANC operations against South Africa in return for the ending of South African support for Renamo, the anti-socialist resistance movement in Mozambique. It was generally assumed that the reason for the CMEA rejection of Mozambique's applications was primarily the cost of taking on another member which would expect substantial aid. By 1981 all the European CMEA countries were experiencing a serious decline in their economic growth rates. The opposition was probably strongest from the smaller East European countries, but the Soviet Union must also have felt that with its own economic problems, including continuing bad harvests and a world oil market starting to weaken, this was a new burden which it could not afford to shoulder.

If this analysis is correct, it is unlikely that there will be any further additions to the membership of the CMEA for some considerable time, despite the existence of a number of communist countries in the Third World which generally support Moscow politically. Another reason, in addition to the cost of the expected aid, is that, with one obvious exception—Afghanistan—they are all geographically remote from the USSR–East European core. To the extent that the interpretation of the long-term aims of CMEA as integration into what would effectively be a single planned economy is correct, non-contiguous members are a complicating factor and enthusiasm for them may be limited for that reason.

The CMEA itself

This chapter has essentially been devoted to setting the scene. Brief surveys of the growth and trade record and the economic system in each CMEA member country have been intended to give the reader an outline sketch of the present state of affairs, as a prelude to the discussions of problems in Chapter 7 and reform experience and prospects in Chapter 8. The preparatory description is now completed with a brief account of the development of the Council for Mutual Economic Assistance itself including indications of some major difficulties encountered. This will be followed up in Chapter 9 by further discussion of the problems facing the organisation and possible developments in reaction to its recent experiences.

The fact that the CMEA did very little during the early years after its foundation in 1949 may be seen as less paradoxical than the fact that it was founded at all at a time when all the smaller member countries were embarking on a strategy of autarkic economic development modelled on that pursued by the Soviet Union since 1930. The Soviet Union had consciously aimed at self-sufficiency and independence from the world market and initially the individual CMEA countries essentially did the same, planning the development of as wide a range of industries as possible in each country. With domestic policies of this kind there was relatively little for the CMEA to do, though the lack of raw materials in many of the smaller countries necessitated trade with the Soviet Union. This meant that they had to face a problem which had not been very important for the Soviet Union, namely the basic difficulty of reconciling centralised planning of production and distribution with foreign trade. The Soviet attitude was to treat the external market as one to which the country turned in order to plug vital gaps in domestic supplies, financing the purchases by exporting whatever could be spared and could be sold abroad; the fact that the Soviet Union was a gold producer facilitated this approach. But for countries whose foreign trade was much greater in relation to their gross domestic product it could not be satisfactory, and a system of bilateral trade and payments agreements developed between CMEA members. This provided a solution to the basic difficulty of reconciling external trade with internal central planning by bringing trade into the planning process. Pairs of CMEA countries agreed in advance what goods they would trade, so that these supplies and uses could be incorporated in their annual plans.

The basic function of the CMEA, then, was and still is to provide a framework under which these bilateral trade agreements are worked out, permitting the combination of central planning with extensive foreign trade involvement. The CMEA is thus not a supranational body but a forum for co-ordination between nations, albeit one in which the Soviet Union is generally presumed to be the strongest influence, owing to its disproportionate economic weight as well as its political position. Beside the basic work of planning intra-CMEA trade, the council has also developed a wide range of subsidiary agencies for co-ordination and co-operation in specialised fields.

Although trade between member countries grew quickly under this system, by the mid-1960s the more developed smaller countries were becoming concerned about the efficiency of their foreign trade operations. One indication of this was an extensive discussion in the economic literature, most notably in Poland, Hungary, Czechoslovakia and East Germany, of criteria for deciding whether a particular export or import transaction was advantageous for the country. Sophisticated formulae and even computer models were put forward, but it is doubtful if they had any great influence on actual trade, which continued to be planned as before.

Another subject of concern was the bilateral nature of trade, which was felt increasingly to be restricting the development of efficient specialisation by the smaller countries. This concern led to the establishment of the International Bank for Economic Co-operation (IBEC) in 1964. The arrangement whereby intra-CMEA trade is conducted in transferable roubles and the IBEC operates a clearing system for member countries' trade payments was supposed to facilitate more mulitilateral trade. In this it has achieved only very limited success. The reason is that unless such trade is planned in advance, the whole CMEA trading system militates against it. Whilst some specific multilateral deals are reached, full multilateralism appears incompatible with planning but is something which results from individual trading decisions when currencies are convertible and spendable. CMEA agreements aim at balanced trade, and a trade surplus is of no benefit; rather, it involves a cost, because what it means is that goods have been sent abroad and real payment may not be received until the next year's trade agreement plans a compensating net flow in the reverse direction. Interest on such credit balances is unattractive. The surplus cannot, of course, be converted into another currency and usually it cannot be spent immediately even in the country in which it was earned because the materials allocation system which is part of central planning means that most goods are not available for purchase without a prior allocation entitlement. This is what is known as goods inconvertibility, as distinct from inconvertibility into another currency.[3]

Goods inconvertibility is a fundamental feature, stemming directly from the practice of central planning of production and distribution. It is what necessitates the CMEA system of bilateral trade agreements and obstructs multilateralism within the CMEA. (It does not, of course, affect trade between a CMEA member and an outside country—unless that country also has a similar domestic economic system.) The unwillingness to tie up resources in a surplus with another member country has increased as growth rates have declined and shortages of many goods become more severe. This has led to the distinction between hard and soft goods in CMEA trade. Hard goods are those which can be readily sold for convertible currency outside the CMEA area or are at least in demand in all CMEA countries: fuels, raw materials and meat are examples. In this situation there is growing reluctance to trade hard goods for soft and a tendency to aim for what has become known as structural bilateralism, or balanced trade not only in total but also within commodity groups. Obviously these attitudes are inimical to economically efficient trade and specialisation.

One reaction to the hard and soft goods phenomenon has been the growth of intra-CMEA trade settled in convertible currency, which is now estimated to account for as much as 10 per cent of turnover in some cases. This overcomes reluctance to sell hard goods within the CMEA but is logically contrary to the purpose of promoting socialist economic integration through a CMEA system insulated from the world capitalist market. One cause of the problem is evidently the failure of CMEA prices to reflect true scarcities; in essence, hard goods are hard because they are underpriced in relation to demand and supply, and if all prices were at equilibrium levels all goods should be equally hard (or soft). This is one reason why Hungary, in particular, advocates alignment of domestic and world market prices. Hardness is also a function of saleability on the world market: homogenous goods which can always be sold for convertible currencies by undercutting other suppliers' prices are regarded as 'hard', whereas many CMEA manufactures, being differentiated products, cannot be sold in significant quantities outside the CMEA at any price because of the quality and design weaknesses, and are therefore 'soft'.

A second bank, the International Investment Bank, was established within the framework of the CMEA in 1971. Its purpose is to promote longer-term economic co-operation and co-ordination among the member countries by joint financing of and/or joint participation in key investment projects. The resources of the bank, although significant, are not very large and, most important, they are purely financial. Consequently the possession of an IIB loan does not by itself permit an investment project to proceed; it is also necessary to negotiate the relevant trade agreements with countries which are to supply equipment or materials before the loan can be transformed into physical resources. This, of course, is just another instance of goods inconvertibility, and where scarce equipment and materials are concerned the availability of finance is usually the least of the difficulties.

To some extent, at any rate, because of this weakness in the conception of the IIB, the emphasis in the investment sphere in CMEA was concentrated from the mid-1970s on major investment projects with joint participation by members. In 1975 the Agreed Plan for Multilateral Integration Measures (APMIM) was adopted, to apply for the 1976–80 period. The Agreed Plan covered a collection of ten large investment projects, eight in the Soviet Union and most already started. Of their total cost of 9 billion roubles, 6.5 billions' worth were the subject of joint financing, with the East European countries responsible for 3.4 billion roubles. The projects were primarily aimed at developing Soviet energy and raw materials production and transport capacity. This APMIM was to be integrated into the participating countries' national plans for 1976–80, so as to ensure the production of the necessary goods, delivery of which was promised in foreign trade plans. A second APMIM, for 1981–5, was not adopted until 1981; it amounted to a detailed scheme of implementation during those years of the Long-Term Special-Purpose Programmes for Co-operation',[4] more often called simply target programmes.

There are five of these target programmes; the first three, adopted in 1978, relate to energy and raw materials, engineering-industry specialisation, and food supply. In 1979 a further two programmes were adopted, for industrial consumer goods and the CMEA regional transport system. The main emphasis has been on the first programme, which included and expanded the first APMIM, adding a set of projects for the development of nuclear energy. The engineering programme was seen as a supporting basis for the energy programme to start with, and later for the other three too. The transport programme also was meant initially to facilitate the movement of heavy freight, mainly raw materials, between the USSR and its partners. The remaining two programmes seem to have received much less attention. The programmes involve some 232 multilateral agreements, of which 120 were reported signed by January 1981; the key ones are the general joint investment agreements. These are supplemented by a set of bilateral arrangements between the Soviet Union and the other member countries, formulating the East Europeans' capital goods contribution (and, in the case of the Orenburg pipeline, labour) and the conditions of Soviet deliveries after completion of the project.

Particular attention has been paid here to the joint investment projects both because these have been the main direction of CMEA activity in recent years and because they also appear to indicate a trend which will continue. The question of national sovereignty over economic development policy within CMEA has long been a touchy one. In the early 1960s Khrushchev advocated some kind of supranational planning and encountered strong opposition, not only from

Romania, though that case is the best known. The 1971 Comprehensive Programme was interpreted by some commentators as presaging the same sort of development, though others disagreed; this intention doubtless underlay the Programme, though opposition continued and its immediate, visible impact was small. What has happened is that the smaller countries' energy and raw materials dependence has given the Soviet Union the opportunity to dictate the terms on which continued supplies will be made available. The emphasis in CMEA has swung away from the earlier discussion of the need to facilitate multilateralism by altering the settlement arrangements. Instead, the general trend now is toward production integration and planned interdependence, and the combined economic and political power of the USSR enables it to impose this, its preferred path of development, on the smaller member countries, though Romania and Hungary, which have geographically diversified their trade most widely, are to some extent exceptions, Romania as part of its generally greater political independence from the Soviet Union and Hungary because of its differing economic system.

Notes

1. *Economic Bulletin for Europe*, 1980, Geneva, cited in Jan Winiecki, 'Are Soviet-type economies entering an era of long-term decline?', *Soviet Studies*, vol. 38, no 3, July 1986, pp. 325–48.
2. The whole of this section is heavily indebted to P. J. D. Wiles (ed.), *The New Communist Third World*, Croom Helm, London, 1982.
3. As far as I know this term was first used, and the importance of the phenomenon stressed in P. J. D. Wiles, 'On Purely Financial Convertibility', *Banking, Money and Credit in Eastern Europe*, NATO Colloquium, Brussels, 1976, pp. 119–26.
4. On these, see, for example, Marie Lavigne, 'The Soviet Union inside Comecon', *Soviet Studies*, vol. 35, no. 2, April 1983, pp. 135–53.

Table 6.1. Annual Growth of National Income (% increase on previous year)

	Soviet Union	Bulgaria	Czecho-slovakia	East Germany	Hungary	Poland	Romania	Mongolia	Cuba	Vietnam
1970	8.9	7.1	5.5	5.5	4.9	5.2	6.7	8.7	—	—
1971	6.0	7.1	4.5	4.5	6.5	8.3	13.9	4.0	—	—
1972	3.8	7.6	5.6	5.6	5.0	10.3	9.8	4.8	—	—
1973	9.1	8.3	4.7	5.9	7.4	11.2	10.7	9.2	—	—
1974	5.0	7.3	6.2	6.1	6.9	10.5	12.6	9.2	—	—
1975	4.8	8.9	6.3	4.7	5.0	9.5	9.8	6.2	—	—
1976	5.3	6.5	4.0	3.5	3.0	7.0	9.9	5.8	1.0*	15.0
1977	5.0	6.4	4.3	5.3	8.3	4.9	8.8	1.4	4.0*	1.7
1978	4.8	5.5	4.1	3.7	4.2	3.1	7.4	8.8	11.4*	2.6
1979	2.6	6.6	3.1	3.5	2.2	−2.4	6.1	8.7	4.3*	−0.8
1980	3.8	5.7	3.0	4.2	−0.8	−4.0	2.5	3.4	3.3*	−5.0
1981	3.1	5.1	0.0	5.0	2.6	−11.8	2.5	8.3	27.4*	8.0
1982	4.2	4.4	0.0	2.4	4.5	5.4	2.0	8.7	3.8*	8.2
1983	4.0	2.8	2.5	4.7	1.2	5.7	4.0	6.1	5.5*	9.1

* Gross social product, in current prices.
Source: CMEA Statistical Annual, various years.

Table 6.2. Annual Growth of Gross Industrial Production (% increase on previous year)

	Soviet Union	Bulgaria	Czecho-slovakia	East Germany	Hungary	Poland	Romania	Mongolia	Cuba	Vietnam
1970	8.9	9.7	9.1	6.5	7.2	8.3	12.1	—	—	—
1971	8.0	8.9	6.7	5.5	6.7	8.1	11.7	9.0	—	—
1972	6.5	9.1	6.8	6.3	4.8	10.4	11.8	12.8	—	—
1973	7.0	9.2	6.4	6.9	7.4	11.3	14.6	8.9	—	—
1974	8.1	8.2	6.5	7.3	8.2	11.5	14.6	8.2	—	—
1975	7.5	9.5	6.9	6.4	4.8	11.2	12.0	6.9	—	—
1976	4.8	6.8	6.5	5.9	4.6	9.3	11.5	7.7	—	13.0
1977	5.7	6.8	5.6	4.8	6.6	7.5	12.7	3.6	—	9.7
1978	4.8	6.9	5.0	4.7	4.9	4.8	9.0	6.9	—	5.6
1979	3.4	5.4	3.7	4.5	3.0	2.1	8.2	14.1	—	−4.6
1980	3.6	4.2	3.5	4.7	−1.8	0.0	6.5	10.2	2.5	−9.6
1981	3.4	4.9	2.0	4.7	2.9	−10.8	2.5*	10.4	23.3	15.0
1982	2.9	4.6	1.1	3.2	2.2	−2.1	1.1*	10.3	4.0	13.8
1983	4.2	3.9	2.8	4.1	0.7	5.9	—	9.6	5.3	14.9

* Commodity production.
Source: *CMEA Statistical Annual*, various years.

Table 6.3. Annual Growth of Gross Agricultural Output (% increase on previous year)

	Soviet Union	Bulgaria	Czecho-slovakia	East Germany	Hungary	Poland	Romania	Mongolia	Cuba	Vietnam
1970	8.6	4.5	0.8	3.7	-5.4	2.4	-5.3	—	—	—
1971	1.0	1.4	3.3	0.0	9.0	3.2	19.4	-1.0	—	—
1972	-4.0	5.7	3.9	10.7	3.0	8.5	9.5	-5.1	—	—
1973	16.5	1.3	4.5	-0.8	6.6	7.8	0.6	13.8	—	—
1974	-3.5	-1.3	2.2	7.3	3.4	1.3	1.2	6.5	—	—
1975	-5.5	7.4	-1.4	-3.0	-1.3	-1.9	3.0	7.9	—	—
1976	5.8	4.4	-2.2	-4.7	-2.7	-1.4	17.1	-8.9	3.6	10.0
1977	4.6	-4.8	9.6	6.5	10.7	1.3	-0.5	-6.3	4.3	-4.5
1978	2.6	4.4	2.0	2.2	1.8	2.7	2.5	16.2	5.8	0.0
1979	-4.3	6.0	-4.0	2.2	-0.6	-1.4	5.4	1.6	1.6	6.7
1980	-3.0	-4.7	5.9	0.0	3.7	-10.7	-4.1	-12.9	-1.6	6.3
1981	-1.0	5.8	-2.5	1.6	2.1	3.7	-1.2	12.0	14.3	2.5
1982	5.5	5.5	5.0	-4.0	7.6	-2.7	8.0	10.7	-5.6	8.2
1983	6.0	-7.4	4.0	4.2	-3.2	3.7	-1.7	1.5	-1.5	3.0

Source: CMEA Statistical Annual, various years.

110 ECONOMIC DEVELOPMENT

Table 6.4. Annual Growth of Total Volume of Capital Investment (% increase on previous year)

	Soviet Union	Bulgaria	Czecho-slovakia	East Germany	Hungary	Poland	Romania	Mongolia	Cuba	Vietnam
1970	11.4	10.5	5.4	7.4	13.8	4.5	11.5	—	—	—
1971	7.3	1.9	5.1	0.5	11.3	7.3	10.7	8.0	—	—
1972	7.2	9.7	8.5	3.9	-1.3	23.6	10.3	0.0	—	—
1973	4.7	7.1	9.5	8.0	3.7	25.0	8.2	7.4	—	—
1974	7.1	7.6	9.2	4.3	9.1	22.4	13.3	29.3	—	—
1975	8.7	17.4	8.4	3.8	13.4	14.4	15.2	18.0	—	—
1976	4.0	1.0	4.0	7.0	0.0	1.0	9.0	37.0	12.0	32.0
1977	4.0	14.0	3.0	6.0	13.0	3.0	12.0	17.0	7.0	22.0
1978	6.0	0.6	4.0	3.0	5.0	2.0	16.0	24.0	-5.0	6.0
1979	0.7	-2.0	2.0	1.0	1.0	-8.0	4.0	-30.0	-1.0	-5.0
1980	2.0	8.0	2.0	0.0	-6.0	-12.0	3.0	19.0	5.0	-2.0
1981	4.0	10.0	-5.0	3.0	-5.0	-22.0	-7.0	32.0	17.0	-18.0
1982	4.0	4.0	-2.0	-5.0	-2.0	-12.0	-3.0	-3.0	-6.0	-6.0
1983	6.0	1.0	0.3	0.0	-3.0	9.0	2.0	5.0	14.0	—

Source: CMA Statistical Annual, various years.

Table 6.5. Population, Natural Increase and Growth of State Employment

	Soviet Union		Bulgaria		Czechoslovakia		East Germany		Hungary	
Total population at the end of 1983 (in millions)										
	273.8		8.95		15.44		16.70		10.68	

Rates of increase of population and employment (% change on previous year)

	Pop.	Emp.†	Pop.	Emp.	Pop.	Emp.	Pop.	Emp.	Pop.	Emp.
1960	1.78	100	0.97	100	0.67	100	0.34	100	0.45	100
1966	1.11	124	0.72	124	0.64	114	0.30	101	0.24	112
Index (1960 = 100)										
1970		145		155		127		107		126
1970	0.92	2.58	0.72	3.81	0.43	1.98	−0.02	1.15	0.31	1.78
1975	0.88	2.39	0.63	7.36	0.81	1.50	−0.35	1.13	0.60	1.32
1980	0.79	1.72	0.34	−0.4	0.41	0.77	0.04	0.51	0.03	−1.12
1981	0.83	1.30	0.35	1.21	0.37	0.75	0.03	0.50	−0.02	−1.0
1982	0.88	1.05	0.33	1.21	0.35	0.45	0.07	0.58	−0.10	−1.5
1983	0.98	0.77	0.23	0.39	0.28	0.62	0.07	0.51	−0.20	−2.30

	Poland		Romania		Mongolia		Cuba		Vietnam	
Total population at the end of 1983 (in millions)										
	36.75		22.53*		1.82		9.95		57.99	

Rates of increase of population and employment (% change on previous year)

	Pop.	Emp.	Pop.	Emp.	Pop.	Emp.	Pop.	Emp.†	Pop.	Emp.†
1960	1.50	100	1.04	100	3.27	100	2.40	—	—	—
1966	1.00	118	0.60	133	—	120	2.72	—	—	—
Index (1960 = 100)										
1970		139		157		139				
1970	0.85	1.23	1.16	3.05	2.79	—	2.14	—	—	—
1975	1.02	2.33	1.04	4.58	2.94	6.96	1.53	3.60	—	—
1980	0.96	0.34	0.76	2.18	2.75	4.60	0.84	−1.49	—	1.19
1981	0.97	−1.72	0.70	0.98	2.75	5.57	0.81	1.27	—	0.42
1982	1.02	−2.14	0.53	1.26	2.72	3.23	1.05	9.19	2.26	—
1983	1.02	0.21	—	—	2.64	5.11	1.08	5.98	2.30	5.05‡

Notes: *End of 1982.
† Employment: Cuba and Vietnam exclude private sector; Soviet Union excludes collective-farm members.
‡ Increase from 1981 to 1983

Table 6.6. Growth of Foreign Trade Turnover (in current prices) (% increase on previous years)

	Soviet Union	Bulgaria	Czecho-slovakia	East Germany	Hungary	Poland	Romania	Mongolia	Cuba
1971	7.0	12.0	9.0	7.0	14.0	11.0	10.0	17.0	−5.0
1972	10.3	11.6	8.3	10.3	7.9	18.9	14.5	13.7	−12.6
1973	20.3	13.6	11.9	14.4	13.8	25.8	24.6	23.3	33.7
1974	26.1	23.9	19.7	20.0	23.6	31.3	38.9	11.0	70.3
1975	28.5	23.9	14.6	16.0	16.2	21.6	6.4	3.8	36.0
1976	11.7	8.7	11.0	14.9	13.4	9.1	14.7	14.8	−5.1
1977	11.7	13.5	11.9	7.4	16.7	8.0	14.7	11.5	10.2
1978	10.8	11.5	8.9	5.6	12.8	7.1	11.1	6.6	10.4
1979	14.5	11.7	10.6	12.2	13.3	9.3	19.2	15.5	2.7
1980	17.2	14.3	10.9	10.3	13.4	5.5	22.4	15.8	17.7
1981	16.6	15.3	7.5	10.7	9.1	−12.4	1.9	23.4	8.7
1982	9.0	10.3	9.1	9.2	9.2	−0.4	−14.0	15.3	12.2
1983	6.6	8.8	9.0	10.6	14.0	9.7	−2.3	13.8	12.2
Index (1970 = 100)									
1960	46	31	50	47	38	39	36	—	—
1970	100	100	100	100	100	100	100	100	100
1980	426	383	300	303	386	385	494	345	364
1983	577	530	384	405	524	369	411	558	498

Table 6.7. Geographical Distribution of CMEA Countries' Foreign Trade (in % shares) (in current prices)

	1970		1975		1980		1983	
	Exports	Imports	Exports	Imports	Exports	Imports	Exports	Imports
Soviet Union								
Socialist countries	65.4	65.1	60.7	52.4	54.2	53.2	55.6	56.5
of which CMEA	55.4	57.0	55.6	48.3	49.0	48.2	50.7	51.7
Developed West	18.7	24.0	25.6	36.4	32.0	35.4	28.9	31.4
LDCs	15.9	10.9	13.7	11.2	13.8	11.4	15.5	12.1
Bulgaria								
Socialist countries	79.3	76.2	80.0	72.3	70.8	78.9	76.4	79.8
of which Soviet Union					50.1	57.2		
Developed West	14.2	19.1	9.3	23.6	15.8	17.2	10.5	13.9
LDCs	6.5	4.7	10.7	4.1	13.4	3.9	13.1	6.3
Czechoslovakia								
Socialist countries	70.6	69.4	71.6	69.8	69.6	70.2	74.6	79.2
of which Soviet Union					35.7	36.0		
Developed West	20.4	24.5	19.8	24.6	21.7	24.3	16.4	16.7
LDCs	9.0	6.1	8.6	5.6	8.7	5.5	9.0	4.1
East Germany								
Socialist countries	73.9	69.4	73.2	66.6	68.7	63.3	63.6	66.1
of which Soviet Union					34.9*	34.9*		
Developed West	21.9	26.7	22.4	29.0	24.1	30.5	29.9	28.9
LDCs	4.2	3.9	4.4	7.2	6.2	6.5	6.5	5.0

Table 6.7. (cont.)

	1970 Exports	1970 Imports	1975 Exports	1975 Imports	1980 Exports	1980 Imports	1983 Exports	1983 Imports
Hungary								
Socialist countries	65.6	65.0	72.2	66.2	55.1	51.1	54.3	52.8
of which Soviet Union					29.3	27.7		
Developed West	28.0	28.8	21.4	27.0	33.9	39.4	32.6	34.2
LDCs	6.4	6.2	6.4	6.8	11.0	9.5	13.1	13.0
Poland								
Socialist countries	63.9	68.6	59.9	45.8	55.9	55.6	67.2	75.2
of which Soviet Union					31.2	33.2		
Developed West	28.4	25.8	31.5	49.3	34.4	35.0	23.5	19.9
LDCs	7.7	5.6	8.6	4.9	9.7	9.4	9.3	4.9
Romania								
Socialist countries	58.1	53.9	46.0	34.5	44.3	38.2	48.0	58.7
of which Soviet Union					19.0	15.8		
Developed West	31.9	39.5	31.5	41.9	35.5	31.2	52.0	41.3
LDCs	10.0	6.6	22.5	14.6	20.2	30.5		
Mongolia								
Socialist countries	—	—	—	—	98.4	98.5	99.4	98.5
Developed West	—	—	—	—	1.6	1.5	0.6	1.5
Cuba								
Socialist countries	—	—	—	—	—	—	86.1	86.8
Developed West	—	—	—	—	—	—	8.7	10.1
LDCs	—	—	—	—	—	—	5.2	3.1

* In 1979.

7 The major economic problems facing the CMEA member countries in the 1970s and 1980s

The preceding chapter surveyed the economic development of CMEA member countries individually and presented a range of statistical data illustrating the trends described. This chapter considers some of the principal economic problems confronting these countries. Whilst there are, of course, variations between individual countries, there is a very substantial measure of similarity between the major problems faced by all the European CMEA members. Consequently they are discussed as a group, focusing on particular problems in turn, and referring to individual cases, and exceptions, as appropriate.

These problems have implications for the activities of the CMEA itself as an international institution aiming to promote economic integration between its member countries. The implications were touched on in Chapter 5 and will be dealt with in greater detail in Chapter 8.

A few words of comment on the treatment of the European CMEA countries as a group seem appropriate here: this is a question on which there has been disagreement among scholars, some regarding all the smaller East European countries as virtually carbon copies of the Soviet Union and others insisting that each must always be considered individually. It is indisputable that they display a wide variety of historical, cultural, political and economic backgrounds, as well as differences in resource endowment. It is equally clear that there must be some significant differences between a country as large as the Soviet Union and all the other, very much smaller, ones. But all the smaller countries have now experienced basically the same system of central economic planning for some thirty-five years (though there have been significant, but not truly fundamental, differences in Hungary since the 1968 reform). In the Soviet Union, the originator of the system, it has been in operation for fifty-five years. This means that nobody now working there has experience of working in the country under a different kind of economic system, and the numbers in the smaller East European countries who have such experience are now small and will soon all be beyond retirement age. This lack of people with experience of other economic systems has implications to which attention will be given when discussing reform in Chapter 8. What is of most concern here is the high degree of uniformity of economic system in all the countries of the group: because most of the problems facing them are common to them all, and are basically related to the nature of their economic system, it is reasonable to treat them as a group for the purpose of the present survey.

Declining growth rates

The most striking fact which emerges from tables 6.1 and 6.2 is the decline in the growth of national income and industrial production in the second half of the 1970s. Whilst both the actual rates of growth recorded and the relative degree of decline between the first and second halves of the decade vary from country to country, in all cases growth was slower in 1976–80 than in 1971–5. In Poland, of course, the fall was dramatically worse than elsewhere, and in Bulgaria and Romania growth generally held up better than in the remaining countries. But despite these individual differences, the trend to slower output growth was universal and incontestable.

This deceleration of growth is not seriously disputed by the authorities or economists in the CMEA countries; indeed, there has been much discussion there of the causes for it and of the measures needed to remedy the situation. Two of the basic reasons are illustrated in Tables 6.4 and 6.5. The picture of the trends in factor inputs presented here is more diverse than in the case of output, but the overall impression is again clearly one of declining growth. Once more the Polish figures for capital investment are the most striking, with tremendous increases in 1972–5 contrasting with sharp falls by the end of the decade. In the other countries the trend is much less dramatic, and in some of them big annual fluctuations make it less conspicuous, but all now show rates of increase in investment which are unprecedently low by the standards they set in the early 1970s, let alone the previous two decades. In the case of the Soviet Union, at least, there has been considerable discussion, both among Soviet economists themselves in Soviet learned journals, and among Western observers, of the reliability of the published investment statistics.[1] The question is complicated and the answer far from clear or unanimous, but the basic point at issue is that the costs of creating new productive capacity, both of construction work and of machinery and equipment for installation, have been rising substantially. These cost increases are not, or not fully, reflected in Soviet price indices or in investment data which are supposed to be in constant prices. The published investment statistics thus overstate the real rate of addition to productive capacity, and if the rate of cost inflation is as high as some Soviet economists maintain, there may have been no real increase in investment at all in some recent years.

There is a two-way relationship, of course, between the growth of investment and the growth of output. With the falling output growth rates, investment momentum could only be maintained by raising the proportion of investment in relation to consumption, or by importing capital (which the CMEA countries have done—Poland, again, most of all). Yet once investment is growing more slowly, output growth is bound ultimately to slacken, too, unless productivity can be improved, that is to say, unless the reduced investment can be applied in ways which bring a higher return than before.

Where labour is concerned, the growth strategy of the Soviet Union, which the smaller CMEA countries followed, has always involved drawing an increasing proportion of the total population of working age into employment in sectors other than agriculture. The participation rate (the proportion of people of working age actually in employment) has risen to a level which is extremely high by international standards and cannot now be increased any more since almost all people of working age who are capable of working do so already.

As far as transfer of labour from agriculture is concerned, there does appear to be considerable scope remaining for this in the longer term, as the proportion of employment in agriculture is still high, compared with Western Europe, even in East Germany and Czechoslovakia, the most advanced CMEA countries (Table 7.1). But there is not now surplus labour in agriculture which can be drawn off without heavy investment to replace it. Indeed, there is a shortage of skilled labour in agriculture in many CMEA countries, especially the Soviet Union, and there is even insufficient unskilled labour at peak periods, so that much disruption is suffered in other sectors of the economy because of the temporary despatch of workers (as well as students and soldiers) to help with the harvest. Furthermore, it would not simply be a matter of additional investment in more capital equipment to mechanise jobs now done manually, thus relieving labour for other sectors of the economy. The quality, above all the reliability, of equipment would have to be greatly improved, as would servicing and repair facilities and spare parts supplies. A capital-intensive agriculture is much more

Table 7.1. Structure of employment, 1983

	Soviet Union	Bulgaria	Czechoslovakia	East Germany	Hungary	Poland	Romania
Industry and construction	38.3	45.0	47.4	50.6	39.2	37.2	44.2
Agriculture and forestry	19.8	22.2	13.6	10.4	23.4‡	30.3	29.0
Other†	41.9	32.8	39.0	39.0	37.4	32.5	26.8

Notes: * Figures are annual averages except for Romania, where they are for 31 December 1982.
† Transport and communication, trade, material and technical supply, services, procurements and the non-productive sector.
‡ Including water management.
Source: *Statisticheskii ezhegodnik stran-chlenov SEV, 1984*, Moscow, 1984. pp. 363–6.

interrelated with and dependent on the rest of the economy than is a labour-intensive one, and when even such elementary requirements as a reliable electricity supply cannot be taken for granted, as in many rural areas they cannot, farms try to hold on to their labour as a precaution even when they have, for instance, electric milking machines, so that cows can, if need be, still be milked manually. Further reduction in the agricultural labour force will doubtless take place, but this is likely to be a slow and gradual process, not one which offers a quick or cheap source of extra labour for industry.

The two major reservoirs of labour are thus now effectively drained dry. Future increments in employment will have to come mainly from population increase, and, as Table 6.5 shows, there is no prospect of salvation from this source. In most CMEA countries the rate of growth of the population was already relatively low by the second half of the 1960s (especially in East Germany and Hungary) and has subsequently declined further. This trend was caused by falling birth rates. Those born in 1966 started to reach working age in 1981 (though the proportion of the age group which will continue into higher and further education and training is probably rising). The possible growth of employment in the 1980s in most cases is severely constrained in comparison with the relative labour abundance on which the earlier rapid economic growth was based.

The need to raise productivity

The East European terminology for this problem is 'the necessity of transition from extensive to intensive development', and it has been widely discussed, and the need to achieve the transition emphasised, for the past decade and a half. Essentially it means that since the scope for achieving growth of output by using more and more inputs of the physical factors of production is now exhausted, future growth must come from increasing the productivity of those factors, and particularly that of labour. East European formulations of this, particularly official ones, are generally vague and speak only of 'productivity'. The experience of market economies is that successful growth may be based on falling capital productivity more than offset by rapidly rising labour productivity, so that increasing total factor productivity is the principle source of growth of output. It is labour productivity that is crucial, both for improvement of living standards and for generation of capital to invest.

It may be objected here that in West European market economies the growth of labour productivity and real wages in the first half of the 1980s has been accompanied by a steep rise in unemployment and comparatively slow growth. This is true, and contrasts with the situation in the United States where there has been a much smaller rate of increase in labour productivity and real wages, but many more new jobs have been created, and growth has been more rapid. However, one major reason for the rise in unemployment in Western Europe has been an increase in the proportion of the working-age population seeking jobs, the rise in the participation rate which had already occurred earlier in Eastern Europe. Western Europe's recent experience is not therefore relevant to most Eastern European countries, where the problem is not to find jobs for people but to employ those people more productively so that their real wages can be raised closer to the level to which they aspire.

What are the reasons for the low level of labour productivity in the CMEA countries and their persistent inability to raise it? The proximate causes may be summarised as technological backwardness and overmanning. Technological backwardness is obviously to some extent just a consequence of the generally lower level of development of the CMEA countries compared with the advanced West. But it is also widely agreed that the pace of diffusion of new technology in the CMEA countries is extremely slow and that there is a strong resistance to innovation inherent in their economic system. Levels of manning, too, are often high even in relation to the technology in use. A striking illustration of this was reported a considerable number of years ago now by the Soviet economist, Manevich, who examined the manning levels at six chemical plants which had been designed in the West.[2] He found that the Soviet economic plans for the plants greatly increased the norms for the number of employees the Western designers had specified, and that by the time the plants were in operation they were actually employing considerably more people than even the Soviet plans envisaged.

Another topic of growing concern in the CMEA countries since the second half of the 1970s, which may appropriately be mentioned at this point, is the increasing scarcity and cost of energy and raw materials. The six smaller East European countries are generally speaking poor in energy and raw materials (except that Poland is an important producer of coal) and thus depend heavily on imports, which have hitherto mostly come from the Soviet Union. (Romania is an exception here, producing some oil of her own, though not now sufficient to meet the whole of her requirements; she has also, as a matter of policy, diversified her sources of imported energy and raw materials to eliminate the earlier high degree of dependence on the Soviet Union). The Soviet Union did not impose the full increases in world prices immediately after the first 'oil shock' in the autumn of 1973, but insisted on a change in the CMEA pricing formula which had the effect of accelerating the impact of the jump in world energy prices. Previously the CMEA member countries had in principle traded during the period of a five-year plan (these plans began in 1971, 1976, 1981 and so on in all the countries) at the average world prices of the preceeding five-year period. This was changed to a moving five-year average, so that in 1976 trade was at world 1971–5 average prices under both rules but under the revised practice trade in 1977 was now at 1972–6 average prices, and similarly thereafter. This has meant that the prices the smaller East European countries paid to the Soviet Union have followed the trend of world prices with a lag, and that with the static or falling oil price of the most recent years the gap between CMEA and world prices has been greatly reduced. As a result the smaller countries' terms of trade have worsened considerably (except for Poland, because of its coal and food exports). The Soviet Union itself also faces greatly increasing extraction and/or transport costs in order to maintain, let alone increase, its output of many natural resources. It has consequently been calling for investment participation from CMEA countries in developing its remotely located reserves. There is also uncertainty about the extent to which the Soviet Union will or indeed can continue to meet East European requirements. It has already refused to supply ever-increasing amounts under the normal CMEA trading arrangements, and set a ceiling at the level of 1980 deliveries; quantities in excess of this level have to be paid for in convertible currency. The reasons are both the increasing cost and difficulty for the Soviet Union in raising production, and the loss of convertible currency if it has to forgo sales to the West in order to supply CMEA members; oil and gas are by far the largest source of hard-currency earnings for the Soviet Union, and its oil production dropped for the first time ever in 1984.

Yet all their studies show that the specific consumption of energy and raw materials in CMEA countries (including the Soviet Union) is extremely high. A Hungarian economist, L. Csaba, quoted some revealing figures from a study by a Czech economist, R. Vintrová, which are reproduced in Tables 7.2 and 7.3. More recent data calculated by the Polish economist J. Winiecki present a similar picture (Table 7.4). These figures of course reflect differences in industrial

Table 7.2. Energy and steel consumption per unit of Gross Social product, 1970 (Czechoslovakia = 100)

	Energy	Steel
Bulgaria	105	75
Hungary	79	76
East Germany	104	95
Poland	105	91
Soviet Union	100	109
France	55	69
West Germany	69	91
Great Britain	76	67

Source: L Csaba, 'World Economic Changes and the CMEA', *Acta Oeconomica*, vol. 25, nos 1–2, 1980, pp. 93–115.

structure as well as in efficiency of resource use, but the breadth and consistency of the discrepancy between the CMEA and the West, both in respect of the levels of consumption and of the rate of reduction is very marked and is surely evidence of difference in efficiency of use. The high figures which the Czechoslovak study shows for the CMEA countries included (except for Hungary) are also, in the opinion of Vintrova and Csaba, attributable to 'the unjustified high share of

Table 7.3. Changes in steel consumption per unit of national income (yearly increase or decrease in %)

	1971–5	1956–75
Czechoslovakia	−1.5	−0.1
Soviet Union	−0.7	−0.8
East Germany	−4.2	−1.5
Poland	−0.9	+2.6*
Hungary	−1.9	−0.5
France	−7.7	−2.8*
West Germany	−7.4	−3.5
Japan	−6.6	+2.3
United States	−3.7	−2.3

* Averge for 1961–75
Source: as Table 7.2

Table 7.4. Energy and steel intensities

	Energy Intensity in 1979 (in kilogrammes of coal equivalent consumed per 1000 US dollars of GDP)	Steel Intensity in 1980 (in kilogrammes of steel consumed per 1000 US dollars of GDP)
Bulgaria	1464	87
Czechoslovakia	1290	132
East Germany	1356	88
Hungary	1058	88
Poland	1515	135
Soviet Union	1490	135
Unweighted average of these six	1362	111
Austria	603	39
Belguim	618	36
Denmark	502	30
Finland	767	40
France	502	42
West Germany	565	52
Italy	655	79
Norway	1114	38
Sweden	713	44
Switzerland	371	26
United Kingdom	820	38
Unweighted average of these eleven	660	42

Source: Jan Winiecki, 'Are Centrally Planned Economies entering an Era of Long-term Decline?', *Soviet Studies*, Table 1.

material-intensive branches of heavy industry, first of all metallurgy and the heavy chemical industry'. But Csaba also adds that the high consumption is explicable 'last but not least, because of the interestedness in gross product resulting from the economic mechanism'.

Planning, incentives and prices

This raises what are the fundamental causes of the excessive use of both labour and material resources and the resistance to the introduction and diffusion of new technology. The system of planning, managerial incentives and prices in operation in all the CMEA countries (Hungary since 1968 has been to a significant extent an exception) have the effect of encouraging excessive use of factors and materials and obstructing innovation. This applies particularly to management at enterprise level, but it also affects higher levels of the planning hierarchy. The essence of the Soviet-type central planning system is that enterprises are set compulsory targets, explicitly or implicitly, for output; enterprise managers receive substantial bonuses if these targets are fulfilled. Because of the amount of work involved, these targets can only be in aggregated form, that is to say, the aims set for the enterprise cannot be specified in complete detail from above. Aggregated targets are expressed in physical or value units. In the first case, there is a direct advantage in producing, for example, heavier rather than lighter machinery, as a given target can be achieved with less processing work that way—but of course it wastes materials, and may also be more expensive to operate. Furthermore, virtually all prices are fixed by central authorities and are not subject to negotiation between supplier and customer (and the supplier–customer connection is also most often imposed rather than established by negotiation), and most prices are based roughly on costs, often individual enterprise costs, with little or no regard for demand or utility. The consequence of this is that use of more, or more expensive, inputs is frequently to the advantage of the producer in meeting output targets expressed in value. And the customer does not resist, because he has little or no choice of supplier, and in any case does not lose if the higher price of his inputs is built into his own plans and can be passed on.

Since the bonuses which managers receive for plan fulfilment are a substantial addition to their income, their incentive to achieve fulfilment by any means is very strong. To anyone accustomed to thinking in terms of a competitive market economy, such behaviour as that just summarised seems absurd, but in the absence of the sanctions of a competitive market, and with great pressure to fulfil output targets, it has become normal. The same kind of reasoning encourages overmanning; once the enterprise plan includes them, there is no penalty for keeping more workers than are really needed. And the extra labour may be useful when there is a rush to achieve a target or (in the Soviet Union especially) when some workers have to be temporarily released to go and help with the harvest. Managers incur penalties if they fail to meet targets, of which the output targets are the most important. They can hoard labour (and materials) at little or no more cost to themselves, so the prudent manager keeps all the reserves possible.

The resistance to innovation is also explicable in a similar way. The introduction of new technologies, either for new products or new processes, disrupts the established activity of production and threatens the fulfilment of output plans. Innovation always involves uncertainty, in any economy, but in a competitive market the reward for successful innovation can be very large, justifying the risks involved. The threat of losing markets to competitors is also a spur to innovation. In the centrally planned economic system of the CMEA countries there are penalties for unsuccessful innovation, if the result is failure to meet targets, which is likely, but there is no penalty for not attempting to innovate and the reward even for successful innovation is generally no greater than for successfully continuing with the old techniques or products. So the balance of advantage and disadvantge is weighted heavily towards playing safe and sticking to familiar products and processes which minimise risk.

It can therefore be argued that the decline in growth in CMEA countries is ultimately attributable to the fundamental features of their economic system, which is becoming less and less suited to the changing economic conditions under which it is required to operate. Chapter 8 will consider the CMEA countries' attempts to alter the system, which was after all evolved in the Soviet Union in the 1930s, where labour was plentiful and the structure of the economy simple. Meanwhile it is worth examining why the decline in growth rates represents such a problem for the CMEA countries.

Pressure for improvement

There appear to be two principal reasons, the relative importance of which differs somewhat in the Soviet Union and the smaller countries. These reasons are, first, the pressure from the population for higher living standards and, second, the wish of governments, most notably of course the Soviet government, to maintain military power. Poland provides the most overt instances of popular discontent with shortages and poor living standards (and at the other end of the spectrum Hungary has been the most successful in meeting consumer requirements). Romania has also experienced worker protest in recent years. East Germany has the special problem that its citizens compare their own situation with that of the West Germans. The reaction of the Czechoslovak and East German regimes, in particular, to events in Poland after August 1980 suggested considerable nervousness and a strong desire to isolate their countries from the possible spread of similar protest. The Soviet Union, despite increasingly serious shortages and a very large and mounting agricultural subsidy bill, has not risked increasing state retail prices of basic foods since 1962. The substantial rise in meat and butter prices in that year provoked a serious disturbance in the city of Novocherkask. Since then there have been some overt price rises on non-food items, concentrated on what are regarded as luxury goods rather than necessities, accompanied by reiterated declarations of the long-established policy of stable basic food prices.

These are all indications that there is strong pressure for higher living standards, and that the authorities recognise it. It can safely be assumed that they would like to satisfy this pressure for greater consumption, even if they may have certain preferences regarding the structure of consumption which differ from those of consumers themselves. An essential plank in the Communist Party platform has always been the promise of eventually high levels of welfare for all, and the majority of the people are becoming more and more impatient at the failure to deliver the goods. Now the regimes face the prospect of slower growth and less favourable terms of trade in many cases, making improvements in consumption still harder to achieve unless labour productivity can somehow be greatly increased.

In the case of the Soviet Union, its role as a superpower requires it at least to match American military capability, from a smaller economy. It is thus crucially important for it to be able to sustain, as a minimum, an economic growth rate equal to that of the United States; failure to do this in the medium and longer term would mean that its military effort constituted a growing relative burden. Just to say this is to highlight the change which has taken place over the past quarter of a century; in 1960 there was much disagreement among Western scholars about the precise rate of Soviet economic growth, but no disagreement that it was higher than that of the United States. The argument was about when, rather than if, the Soviet Union would catch up. One textbook on the Soviet economy even had a table showing the number of years required for Soviet industrial output to match that of the Americans;[3] with Soviet growth ranging from 7 to 9 per cent per annum and American growth ranging from 2 to 4 per cent per annum, the period varied from a minimum of fourteen to a maximum of thirty-two years, assuming that Soviet industrial production was then 45 per cent of that of the Americans. When the author revised his book in 1974 he no longer included such a comparative table and merely concluded, somewhat agnostically, 'The productivity of [Soviet] resources is still far below that which we have achieved, a situation which can be interpreted either as a demonstration of the inefficiency of the system, or as an indication of potential for further gain relative to the United States'.[4] Since the end of the 1970s, at any rate, the future relative movement of the two economies' growth rates has been an open question, and this has undoubtedly been an extremely unpleasant and worrying new experience for the Soviet leaders.

There are, then, two very important reasons spurring the CMEA countries' governments on to seek ways of reversing the fall in their growth rates. These are reinforced by other less substantial factors like the international prestige of socialism and its standing in the eyes of the Third World. The first general recognition in CMEA countries that all was not well with the Soviet-type economic system came in the mid-1960s, when there was a wave of discussions and some measures were adopted to 'reform' the system. The question of reform will be considered in Chapter 8; here it is enough to say only that little or nothing altered in the 1960s, or later, except in Hungary.

Technology imports and foreign debt

From the end of the 1960s the approach changed to one of trying to overcome technological backwardness by importing knowhow and capital equipment, much of it on credit, from the West. The mere fact that this strategy was pursued, with varying degrees of intensity in different CMEA member countries, for the best part of the decade of the 1970s and that that decade ended with unprecedentedly low growth rates, by itself suggests that the strategy was a failure. The well-known debt problem of Poland reinforces this conclusion. It is true that the conditions obtaining at the start of the 1970s on the world market were much more favourable for this strategy than those which prevailed from 1974 onwards, after the first 'oil shock'. The recession this caused led to a severe contraction of the export markets on which the CMEA countries depended to earn the hard currency to service and ultimately repay their borrowings. The terms of trade moved against most of them, as already mentioned. The majority ended the decade with substantial hard-currency debts and significant debt-service ratios and now face the difficulty of earning the currency to meet these obligations in a harsh world economic environment with growing competition from the newly industrialised countries of the Far East. Poland is much the most dramatic and best-known case, because of the sheer scale of her borrowing to finance capital imports. The Soviet Union's position is the strongest both because it has been more cautious in its borrowing and because of its endowment with oil, gas, gold and other natural resources; the other countries fall between these extremes.

It is hard to identify any great benefits compensating for the debts incurred, though some technological modernisation has undoubtedly taken place. However, what was said above about the obstacles to innovation and technology diffusion posed by the central planning system also raises doubts about the extent of the benefit which can be reaped from capital imports. Furthermore, the process of technology transfer itself takes a considerable time, and meanwhile research and development in the countries of origin (the West) continue. Consequently, what was the most modern technology available at the time a deal was concluded is often no longer so by the time it is installed and working in the purchasing country. The inflexibilities of central planning play a part here too, tending to prolong the period of commissioning of the new plant. All in all, the impact of technology imports does not appear to have been large (which is not to deny that they may have been important in certain specific instances), and the debts accumulated in the process are one more constraint on the CMEA countries in the future.

The particular problem of agriculture

Periodic but recurrent food shortages and queues, above all in Poland and the Soviet Union, are the most dramatic signs of unsatisfactory standards of living and causes of discontent. Elsewhere the situation appears less serious, though not good (except, again, in Hungary, where agricultural policy has offered real incentives to producers, resulting in reliable supplies in the shops and substantial exports, notably of meat, to the Soviet Union, paid for in hard currency; in Bulgaria, too, agriculture has been relatively successful). One general point is that agricultural output, by the nature of farming, is more subject to uncontrollable annual fluctuations than is that of other sectors of an economy, and the comparatively high share of agriculture in national product in all the CMEA countries except East Germany and Czechoslovakia (Table 7.5) means that these fluctuations have more effect on total national product than is the case in Western Europe. Soil and climatic conditions in much of the area, particularly the Soviet Union, aggravate this universal proneness to fluctuating output. Romania has evidently been liable to exceptionally sharp fluctuations (see

Table 7.5. Structure of national income produced, 1983 (in %)

	Soviet Union	Bulgaria	Czechoslovakia	East Germany	Hungary	Poland	Romania
Agriculture and forestry	20.3	16.9	8.5	7.8	18.0	18.4	16.7
Industry	46.4	58.1	62.0	68.5	45.3	50.0	60.8
Other*	33.3	25.0	29.5	23.7	26.7	31.6	22.5

* Construction, transport and communications, trade, material and technical supply services, procurements and other sectors of material production.

Source: Statisticheskii ezhegodnik stran-chlenov SEV 1984, Moscow, 1984 pp. 39–40.

Table 6.3). The variability of the Soviet grain harvest has long been notorious; in 1981 it was so bad that the figure for grain output was not disclosed, for the first time since more or less normal publication of economic statistics resumed in the late 1950s after the years of silence under Stalin. Publication has not resumed since 1981, but it is generally accepted that in no subsequent year has grain production come anywhere near the level demanded in the five-year plan, or that achieved in the best years of the 1970s. The consensus among Western scholars now is that the harvest has been too consistently bad in the past five years to be attributable solely or even principally to bad weather. Soviet grain farming has been stagnating at a level far below that needed to ensure self-sufficiency, so that regular and extensive imports for scarce hard currency have been necessary. This in turn restricts the funds available for other kinds of imports from the West and/or places extra pressure on supplies of exportable raw materials. The Soviet Union is fortunate in being the second largest gold producer in the world, but it has to reckon with the effect of its sales on the price of gold, which in dollar terms at least has been depressed in recent years, following the peak in 1980. The Soviet Union cannot necessarily, therefore, finance grain purchases wholly this way, even if it has physical reserves of gold.

Whilst few if any observers would claim that the institutions of CMEA agriculture were conducive to efficient farming, care should be taken not to lay all the blame for the shortcomings on the collective farm. In Poland, after all, about 75 per cent of agricultural land is in the form of smallholdings, whereas in Hungary and Bulgaria most land is collectivised. In the Soviet Union the legacy of ill-will and indifference created by brutally enforced collectivisation may well be nearly insuperable; one Russian economist has said that in killing the peasant's love of private property (which was an ideological and political objective) his love for the land had been killed, too. The collective farm as an institution is thus seriously handicapped, but the price, investment and supply policies adopted towards this sector must bear much of the responsibility for its poor performance. Until at least the late 1960s the CMEA countries' governments mostly neglected agriculture, giving it low priority in comparison with industry. Investment and equipment supplies were inadequate and product prices low. In Poland, although private farming was tolerated because the attempts to collectivise during the 1950s encountered such fierce resistance that the authorities abandoned them, the regime never renounced the long-term aim of 'socialising' agriculture, so private farmers always felt insecure. They were also unable to obtain the kind of modern tools and equipment needed for efficient small-scale farming; most investment went into the state farms. Until 1971 there were compulsory delivery quotas at unattractive prices. The result of all these policies was stagnation of output, leading to the well-known shortages.

In the Soviet Union investment in agriculture was increased greatly after 1965 and was at a high level throughout the 1970s and early 1980s. Production of many kinds of equipment was stepped up, and product prices have repeatedly been raised. These measures were not without result, but nevertheless, the fact remains that Soviet agriculture cannot produce enough food to meet the population's demand, even though the state budget has to provide extremely large subsidies. Production costs are excessive, to a considerable extent because the quantities of both capital *and* labour employed are now quite high. The virtual absence of a rural road network, as known in Western Europe, makes transport of produce extremely costly and causes a high rate of wastage. The indifference of the work-force is also a crucial element, despite levels of pay which are not now greatly inferior to those of many urban workers (though rural amentities are much worse).

There are also two more general and intractable problems relating to agriculture in Soviet-type planned economies which need to be discussed. The first has already been touched on in connection with consideration of agriculture as a source of additional labour for the rest of the economy. This is that as manual work is mechanised, and as the level of use of chemical fertilisers and pesticides increases, so agriculture becomes more dependent on the industrial sector of the economy. It is then adversely affected if manufactured inputs are of poor quality or equipment is unreliable. Because of the nature of farming as an activity, promptness in carrying out maintenance and repairs to equipment is vital; delays caused by poor availability of spare parts can often have more serious effects in agriculture than elsewhere, because the timing of work in farming is crucial to success. Now distribution, spare parts supply and servicing and repairs are all notoriously weak areas of the Soviet-type planning system. As agriculture is more closely integrated into the other sectors of the economy, it becomes more exposed to those weaknesses. It thus becomes less realistic to regard agriculture as a sector apart, where success could be achieved by appropriate agricultural policies alone. Agricultural policies since the late 1960s have been generally sensible, and much money has certainly been spent on agriculture in the Soviet Union, but the returns have been limited and have diminished as the sector has become increasingly subject to the general weakness of the whole administrative central planning system.

The second problem is similar in that it also involves the fundamental economic system. It is that farming may well be more seriously damaged by remote and inflexible management and bureaucratic interference than any other sector of the economy. The reason is once again the crucial importance of the time when operations are performed in farming, and the fact that no two farms and no two years are identical. On-the-spot initiative, judgement and interest are therefore vital for farms to respond successfully to the ever-varying natural conditions. In the Soviet Union, at least, many farms are arguably too large for efficient management in any case, but even apart from this, the kind of flexibility required is very difficult to achieve in the bureaucratic and hierachical planning system, the essence of which is that instructions are issued from above and executed by managers who are subordinate to both the planners and the party network.

A related problem, which exarcerbates food shortages (and explains the Polish government's attempts to raise retail prices of food in 1970, 1976 and 1980 and the enormous increases imposed by the Jaruzelski regime in 1982) is that for a long time a deliberate policy of stable retail food prices was pursued and indeed proclaimed in all the CMEA countries. At the same time, prices of manufactured consumer goods were relatively high, and rose further during the 1970s as new (or so-called new) products replaced older types. In the Soviet Union it has been explicit official policy to make increases in the state retail prices of 'luxury' goods such as carpets and the more expensive kinds of furniture, with the aim of absorbing some of the large accumulation of purchasing power, while holding the prices of necessities, including all basic foods, unchanged. This policy has distorted the structure of demand, so that the populations have tried to spend an abnormally large proportion of their income on the relatively cheap food, aggravating the shortages. The authorities in this respect fashioned a rod with which they themselves subsequently came to be beaten. Hungary has been tackling the problem sensibly, with increases in food prices and compensating wage rises, designed to leave the average consumer no worse off but to change the relative prices of goods and thus bring about some shift in the structure of demand away from food. The other countries will surely have to follow this example sooner or later. China too, incidentally, has reached the same conclusion.

The core of the CMEA agricultural difficulties is Soviet agriculture, with Poland forming a secondary problem. The Soviet population is more than double that of the six smaller countries together; if the Soviet problem remains unsolved, the other members, however successful, could only provide marginal help in relation to the scale of Soviet demand; on the other hand, if the Soviet Union succeeded in solving its own problem it would then *ipso facto* be in a position to help out the other countries when they suffered poor harvests. At present, despite the undoubted seriousness of Soviet efforts to raise agricultural output over the past twenty years at least, the outlook is probably less promising than it was a decade ago. The more obvious measures— basically increases in the resources put into agriculture—were taken then; they have yielded only limited success. Further improvements— depending on much greater efficiency of equipment and supplies, radical improvements in the rural infrastructure to provide faster, cheaper goods transport and more amenities for residents, and above all a change of attitudes towards agriculture—will be much harder to achieve, and are much more closely tied up with the success or failure of efforts to improve the efficiency of the economic system as a whole.

It is clear, therefore, after surveying the range of problems which all the European CMEA member countries face, if in varying degrees, that they are ultimately all linked with the nature of the system of administrative central planning. This system was evolved in the context of the Soviet Union at the start of its 1930s industrialisation drive, when the objectives and circumstances were very different from those of the 1970s and 1980s. The subject of Chapter 8 will be to examine the attempts which have already been made to reform the CMEA countries' economic system, why they have broadly speaking failed, and what the prospects are for further reform efforts.

Notes

1. See A. Nove, 'A Note on Growth, Investment and Price Indices', *Soviet Studies*, vol. 33, no. 1, January 1981, pp. 142–5; the comments by Stanley H. Cohn, *Soviet Studies* vol. 33, no. 2, April 1981, pp. 296–9; Nove's reply, pp. 300–1; P. J. D. Wiles, 'Soviet Consumption and Investment Prices and the Meaningfulness of Real Investment', *Soviet Studies*, vol. 34, no. 2, April 1982, pp. 289–95; and the Soviet economists V. Krasovsky and V. Fal'tsman whom they cite.
2. E. Manevich, *Voprosy ekonomiki*, no. 10, 1969, pp. 31–7.
3. Robert W. Campbell, *Soviet Economic Power,* Houghton Miflin Company, The Riverside Press, Cambridge, MA., 1960, pp. 192–5.
4. Robert W. Campbell, *Soviet-type Economies*, London, Macmillan, 1974, p. 115.

8 Economic reforms and attempts to improve efficiency: prospects for the future

Neither academic nor official criticism of economic inefficiency in the CMEA countries is purely a recent phenomenon. As early as the beginning of the 1960s (and from the mid-1950s in Poland and Hungary) there was widespread discussion of the need to overhaul the system of central planning which had been evolved in the Soviet Union in the 1930s and introduced in all the smaller East European countries at the end of the 1940s. The system had, of course, been born in a country where the level of development was relatively low and the government attached very high priority to the fastest possible build-up of basic heavy industry. At least two of the countries to which the system was applied after the Second World War, East Germany and Czechoslovakia, already possessed a much more highly developed industrial structure, and Hungary and Poland were also considerably more advanced. By the start of the 1960s there was consequently a broad feeling that the system was no longer as appropriate to the tasks it faced as it might once have been, while many economists in Poland, Hungary, Czechoslovakia and East Germany, at any rate, believed that it had never been appropriate, at least for their countries.

The first wave of reforms

The outcome of this situation was that during the mid-1960s a wave of 'economic reforms' swept through the CMEA countries. The timing and the detail of the changes introduced varied considerably from country to country (and Romania showed rather little interest), but the general themes were the same everywhere. They amounted essentially to some degree of decentralisation of economic decision-making, with reduction in the number of compulsory plan indicators, declarations of intent to increase enterprise autonomy, simplification of the administrative structure of planning and plan implementation, and various financial measures designed to make enterprises somewhat more responsible for earning the resources they required. Prices generally underwent a thorough review, though the principles on which they were set, and the principle of central price-setting, were in most cases unchanged.

Some of the participants in the discussions of economic reform advocated much more far-reaching changes than those we have just summarised. There was much talk of the need for concern with profitability, and indeed most of the reforms introduced used profit in some form as a target or success indicator. This caused quite a lot of confusion and misunderstanding at the time, certainly in the West. It subsequently became clear that not all proposals involving the use of profit as an enterprise target meant the same thing. Some schemes, like the package of reforms introduced by Kosygin in the Soviet Union in 1965, saw profit as a general or synthetic measure of the efficiency with which planned tasks were performed, using allocated inputs to produce the products indicated in the plan, in a situation where all prices continued to be centrally fixed. In this conception there was no question of profit opportunities influencing the structure of output. In contrast to this, the Czechoslovak economists, led by Ota Šik, and the majority of Hungarian economists wanted enterprises freed of output targets and left to maximise profits by producing what was in demand, buying their inputs from the cheapest (domestic) supplier; some degree of price control was needed, at least temporarily, they conceded, but in principle, product prices, they argued, should be negotiable between buyer and seller and reflect market situations.

By the beginning of the 1970s the situation had settled down and it was possible to distinguish two clearly different models of reform which had emerged from the discussions and the changes actually introduced during the 1960s. The first may be called a relaxed and rationalised administratively planned economy. It aimed, basically, to make the traditional system of central planning work more efficiently by devolving detailed decision-making to enterprises and accordingly cutting down the volume of planning work and the number of targets set. It also sought a less cumbersome administrative structure above enterprise level, with fewer organisations and, it was hoped, more clearly defined powers and responsibilities. The granting of more operating autonomy to enterprises within the basic plan framework led to much attention being paid to the reformulation of plan indicators and, above, all, managerial incentive schemes, in an effort to motivate enterprise managers to do what the planners wanted. Prices were also reviewed more frequently than had previously been the practice, but the great majority of prices continued to be administratively fixed. This is the model of 'reform' which was carried out in the Soviet Union, East Germany, Poland, Bulgaria, Romania and, after August 1968, Czechoslovakia.

The second model of economic reform was much more far-reaching. It was based on comprehensive use of market relations, though of course retaining state ownership of productive enterprises and national planning of the basic structure of the economy, with control through a range of parametric measures rather than the issue of direct instructions. It may therefore be called a socialist market economy. Czechoslovakia was preparing to switch to this model during the two or three years before the Soviet invasion in August 1968 but this development was then halted and the country reverted to the administrative planning system. The Hungarian 'New Economic Mechanism', introduced in January 1968, is the only example of the socialist market economy put into practice in the CMEA. East Germany seems to have moved a little way towards it during 1965–8, with enterprises gaining some power to decide their output, but this was reversed after 1968 when what were designated 'structure-determining tasks' were introduced and received heavy emphasis; their effect was to restore the traditional central determination of the product mix.

The crucial features of the Hungarian New Economic Mechanism were the abandonment of compulsory plan targets for enterprises, the dropping of the cumbersome system of allocation of inputs in favour of wholesale trade, and the switch from universal central price-setting to negotiation of prices between buyer and seller as a principle, although a considerable degree of government control over both retail prices and the wholesale prices of some basic materials was retained. Enterprises then decided largely for themselves what products to produce, in the light of what they could most profitably sell, though obviously they were constrained by the set of capital equipment they had at their disposal. Capital investment remained subject to extensive controls, though there was some significant degree of decentralisation here, too.

Hungary has remained the sole case of a socialist market economy in the CMEA (Yugoslavia, of course, is not a member) and the other countries all still adhere in essence to the administrative planning system. Hungary itself reintroduced tighter controls, most notably over the distribution and use of profits, in 1972 and for a time it looked quite possible that the New Economic Mechanism would be abandoned. The initial years of its operation had led to a marked increase in inequality of personal incomes, as successful enterprises paid their employees, particularly senior staff, large bonuses. Producers of agricultural products were also doing well, and many industrial workers became discontented owing to the relative decline in their position. This was used by elements opposed to the new mechanism to enforce a variety of restrictions on enterprises' freedom, but there was no formal return to the setting of output targets or material allocation. While the regulation of wages, prices and investment has continued to involve frequent changes in the relevant rules and taxes, the principle of control by fiscal and monetary parameters has been preserved in the main. It is, however, weakened by the practice of bargaining between individual enterprises and the central authorities over these controls. Since the beginning of the 1980s, Hungary has renewed efforts to move towards a set of prices reflecting world market supply and demand relationships. It has also abolished the ministries responsible for particular sectors of industry and replaced them with a single Ministry of Industry. This was a step which the original designers of the New Economic Mechanism would have liked to take in 1968, but which they deferred in order to avoid arousing the hostility of the vested interests affected. The move is important, because it was the continued existence of the industrial ministries that facilitated some degree of retreat in the direction of the traditional system during the 1970s, when these bodies stepped up their efforts to intervene in the running of 'their' enterprises. It also signifies a restatement of top-level commitment to the principles of the NEM, and both economists and the authorities in Hungary display no doubts that these principles represent the correct line of development, at any rate for their country, although they admit that they expect to remain the exception to the rule in the CMEA for the forseeable future.

Developments during the 1970s

While the rule in CMEA countries is still central administrative planning, the decade of the 1970s witnessed much tinkering with the detailed arrangements and a number of packages of measures which have been graced with the title of 'reform', in or outside the country concerned. One such was announced by the Soviet Union in July 1979. This involved the introduction of a new success indicator to be used 'in the majority of sectors as basic in the planning of production, labour productivity and the wages fund': the indicator was called 'normed net output' and was to be calculated for every product, as a kind of second price alongside the conventional wholesale price.[1] Both, of course, were fixed administratively. Compulsory targets for the volume of output of the most important products, and for increasing labour productivity, continued to be set for enterprises by their superior organs. The decree went into much more detail, but everything it said confirmed the view that no fundamental change would occur or was intended. The measures could not be expected to bring about any serious improvement in the efficiency of the system. The only point which could be singled out for commendation was a change in the targets for construction organisations from volume of work done to buildings *completed*; if this were really enforced, it could help to reduce excessive construction times. Within a few years it was clear that the initial scepticism of Western commentators concerning the measures in the July 1979 decree had been justified. Apart from a temporary revival in growth in 1983 under Andropov's discipline campaign the Soviet economy has shown no major improvement, and reform is now very clearly on the agenda once again under the new leader, Gorbachev.

Another equally superficial set of 'reforms' was proclaimed in Romania in 1978, confusingly called by exactly the same name as the Hungarian reform of ten years earlier, New Economic Mechanism.[2] It was immediately clear from the announcement that this was no change of model but yet another attempt to make the administrative system work in a way that permitted the authorities to implement their wishes more effectively. There was much stress on plan discipline and conformity to regulations and, whilst the indicators used to specify enterprise targets underwent the inevitable modifications, none of the measures was consistent with any increase in enterprise autonomy, despite talk about workers' self-management. The contradiction between this and the much-emphasised 'principle of a unitary national plan' was simply ignored. The economic accountability of enterprises was to be increased, making them rely more on their own funds in implementing expansion plans, but targets for production, investment, exports, imports, home market deliveries and technical progress were retained, and no significant effects on economic efficiency could be expected as a result of such a 'reform'.

In Poland, as mentioned earlier, public proposals to reform the traditional Soviet-type system first appeared following Gomulka's return to power in October 1956. Whilst many Polish economists wanted much more drastic changes, the first result was a modest relaxation of the degree of centralisation in the late 1950s, but this was followed by the tendency towards creeping recentralisation which has since been observed in all cases where the decentralisation stops short of dropping compulsory plan targets for enterprises. Continuing discussions and proposals followed, though little was actually changed, and the last reform prepared under Gomulka's leadership was partly responsible for the discontent which led to his fall. Although it was food-price increases which triggered the disturbances in December 1970, the intended reform would have held down money wages and thus depressed real incomes, so under the new leader, Gierek, it was discarded and a new reform prepared, called the New Economic and Financial System (NEFS). Introduction of this reform commenced in 1973.[3] One of its features was the merger of enterprises into larger associations, a trend which was widespread as an element in the administrative reorganisations which followed the first wave of reforms. For this reason it is also known as the WOG (the Polish acronym for 'large economic organisation') reform. On paper, at any rate, the NEFS envisaged a gradual movement towards something more like the market socialism model in operation in Hungary, though even the proposals in Poland were neither as consistent nor as far-reaching as the changes implemented in Hungary. Enterprises (*Wogi*) were to cease to receive compulsory plan targets and their performance was to be evaluated by their success in raising net profit and value added compared with the preceding year (rather than a target). Prices were revised but not freed from control to any significant extent, except for new products (the consequences of this last provision were disastrous, as will be explained below).

There was, however, a crucial difference between the Polish NEFS and the Hungarian NEM.[4] The latter was introduced throughout the whole economy simultaneously, after careful preparation over the previous two or three years, which included appropriate administrative measures such as the winding up of the supply allocation system and at least a reduction in the number and powers of industrial ministries. The Polish NEFS, in contrast, was to be introduced gradually. Its adoption was optional, at the initiative of *Wogi*. Sometimes it was even known as the 'pilot unit' reform, because the first *Wogi* to operate under the NEFS were regarded as pilot or experimental units. This meant that the new system had to function alongside the old, and all the organisations required for the latter remained in operation, and enterprises continued to be formally subordinated to industrial ministries. The environment was therefore more complicated and potentially hostile, the feeling of total change was lacking, the possibilities of intervention in what was supposed to be enterprises' sphere of authority were much greater, and it was far easier to retreat into the familiar old practices at the first sign of trouble.

The NEFS became widespread, though not universal, by 1975 but by then the economy was in serious difficulty as a result of the enormous boom in investment, rapid rises in wages and soaring foreign debts. The result was that the system became subject to a great number of *ad hoc*, if rather ineffective, interventions and restrictions and never achieved the change in economic model which it might have under more favourable circumstances (though some economists argue that it was always inconsistent and could never have led to a Hungarian type of system).

The pricing arrangements were a particular source of trouble. Prices for existing products remained fixed, and continued to be a bad indicator of demand and supply relationships, but those for new products could be set by producers with little control, and this provision, intended to offer an incentive for innovation by making it more profitable, in fact mainly stimulated the dropping of 'old' products and their replacement by 'new', slightly modified ones so as to permit large price increases. The ability of producers to do this was enhanced by the fact that many were monopolists or near-monopolists as a result of amalgamation into *Wogi*, and in any case sellers' markets prevailed, as is normally the case in centrally planned economies.

Inflation thus became a serious problem. With the central authorities losing control, managers generally found ways of raising wages, too, despite regulations. By the end of the 1970s there was neither a socialist market economy nor a traditional administratively planned economy, but increasing confusion and lack of direction.

The trend towards fewer and larger enterprises was seen in almost all East European countries following the decline of interest in strictly economic kinds of reform at the end of the 1960s (with the exception, of course, of Hungary). It seems to have had two purposes initially. One was to simplify the work of the planning and administrative bodies at the centre, by reducing the number of different organisations with which they had to deal. There probably has been some genuine benefit deriving from the policy in this respect. The second declared purpose was to devolve more decision-making away from the centre, though in our view there has been little effect of this kind.

Reasons for failure of reforms

Economic reform has now been the topic of widespread discussion in CMEA countries for over two decades. Schemes of self-styled reform have been proclaimed, at least one in every country, more in some. Yet only in the case of Hungary has any real change in economic system taken place. The crucial features of the Hungarian reform which brought genuine change there have been explained above. There is now a broad consensus among almost all Western observers and many East European (and even some Soviet) economists that until reforms along similar lines are introduced in these countries, their problems of inefficiency will persist. What must now be considered is why the multifarious detailed changes in the planning systems and economic mechanisms which have been introduced in these countries have been so unsuccessful. Certainly there has been no lack of effort, if effort is measured by the volume of official decrees and pronouncements: the question is why have these not been more effective.

The main points of the answer have already been indirectly suggested by indicating the crucial features that distinguish the Hungarian reform from the others. The first point is that experience shows that compulsory output targets have to be abandoned completely; it is not possible to reduce their number and, so to speak, share the decisions about what to produce between the centre and the enterprises.[5] Attempts to do this always result in a relapse into central planning of output. The reasons are twofold. For one thing, the superior bodies are too strongly accustomed to issuing orders to 'their' enterprises to produce whatever they consider necessary, and cannot or will not alter their ways. They are themselves under pressure from the central organs, the government and the Party to fulfil demanding output plans, and they pass this pressure down the line by the only means they know and have at their disposal. Second, the criteria which could guide enterprises to make appropriate decisions concerning their product mix are lacking: prices are not a reliable indicator of demand. When enterprises are given some freedom to determine their production pattern, and their managers have a profit-based incentive scheme, the result can often be increases in the output of little-needed goods and cuts in production of others which are in very short supply. This is because enterprises do not have to worry about finding markets for their production, and because profit rates vary widely and quite arbitrarily, but prices may not be changed, even when customer enterprises would agree to pay more to get goods they need. This situation thus reinforces the well-established predilection of ministries for issuing production orders; if they do not, urgently required products may not be produced because they are not profitable.

A second point, which emerged very clearly from the comparison of the Hungarian NEM of 1968 with the Polish NEFS of 1973, is the crucial importance of introducing the reforms simultaneously throughout the whole of industry at least.[6] Whenever attempts have been made to operate a new, decentralising reform in part of industry, alongside the traditional system, the latter has always swamped the reform. The reasons are again the continued existence of the old superior organisation with administrative powers over the unreformed enterprises—powers which they inevitably apply in relation to enterprises which are supposed to be autonomous under the reform—and the dependence of the unreformed sectors on inputs from the reformed sector which the latter will not necessarily produce voluntarily.

The notion of economic experiments, whereby reform models are tried out on a small scale first, contains a fundamental flaw in that the economic environment in which such experiments are conducted can only be quite different from the one which would prevail if the reform model were introduced universally. On the one hand, the experimental enterprises are likely to be an unrepresentative selection, however chosen, and will be the subject of special attention and in some respects favourable treatment, particularly as the authorities who have set up the experiment usually want to show that the model concerned is successful. On the other hand, the experimental enterprises face problems because of their interdependence with the sector still on the traditional system and, especially, the limitations of the price system.

It is failure to tackle the fundamental weaknesses of the centralised price system that can be singled out as the final crucial reason why the reform efforts of the 1960s and 1970s have had so little impact. Even though in most cases reforms were accompanied by a price review, in which most if not all prices were revised, the principles on which the new prices were set did not differ greatly from those of earlier price-setting (to be fair, there were some advances, such as inclusion of charges for capital where none had previously been made). More important still, the general principle that prices were centrally determined by a specialised government organ, and were not allowed to be changed even by mutual agreement between supplier and customer, was nowhere varied except in Hungary. This meant that prices remained too inflexible and profit margins too bad a guide to relative scarcities to be usable as signals to tell producers what output was needed, thus undermining attempts to devolve some production decisions to them.

Only in Hungary were the principles of pricing changed; although many controls still remain, the basis was altered to one on which prices were negotiable unless controlled. Controls took the form of actual price-setting in some cases, and in others of the prescription of a range within which negotiation was permissible.

A variety of schemes to use prices to motivate innovation were also unsuccessful; reference has already been made to the inflationary consequences of free pricing of 'new products' in Poland. The basic trouble is that it is not so much newness itself that is desirable but the various improvements that make a new product new. The distinction between new and existing but modified products which must be made for such schemes can only be arbitrary, and once it is advantageous to obtain the classification 'new' producers will be guided by this more than by customers' requirements. Once again, the test of the market place is what is missing. In a seller's market producers, especially if they have a considerable degree of monopoly power, can simply discontinue existing products and replace them with successors which are just sufficiently different to secure classification as 'new', thereby gaining the pricing advantages available for such products, without necessarily offering the customer any improvement but forcing him to pay more.

Obstacles to more effective reforms

The arguments just set out are ones that command a broad measure of agreement among Western and many Eastern scholars. These conclusions had generally been reached in the late 1970s, if not earlier in some cases. Yet in 1985 there is still little sign of these lessons being learnt and applied in practice in the CMEA countries. Even though Gorbachev may proclaim a package of measures labelled as reform, it is highly unlikely that this will involve a change in the fundamental economic model in the Soviet Union. What, then, are the obstacles to more far-reaching reform in the direction of what we earlier designated a socialist market economy? There are many obstacles, ranging from purely economic to basically political and social ones, and this section will survey what seem to us some of the most substantial. The first economic difficulty is that such reforms would have a short-run cost. They would involve considerable structural readjustment, with, temporarily, slower growth of output and possibly some transitional unemployment. Because the administratively planned economy is characterised by chronic shortage, resources would have to be built up first before they could be used in the new directions indicated by market forces. Hungary deliberately created some 'slack' in the economy in the run-up to the introduction of its reform in 1968, and the authorities also made way for the new initiatives which the NEM unleashed, most notably by cutting central investment to accommodate the decentralised investment boom generated by enterprises when they became free to earn and use their funds in response to profit signals. The longer the non-market system has prevailed, the more extensive the necessary adjustments are likely to be, so this cost may well now be greater for the other small countries than it was for Hungary. For the Soviet Union it would probably be greater still. Of course, in the longer run the benefits brought by the reform should outweigh these costs, and few Hungarians would question that this has been the case there, but the short-run consideration is a serious discouragement to regimes facing a range of pressures for more output now.

A further problem is that the scope for economic reform in any one CMEA country is restricted by the intra-CMEA system of trading. The operation of the CMEA is discussed particularly in Chapter 6, but it is relevant here to note that the one important formal exception to Hungarian enterprises' freedom to decide their own output is that they may be required by the central authorities to produce goods needed to meet Hungarian delivery obligations to other CMEA countries. (In practice there is also considerable informal intervention, including pressure to maintain production of exports for convertible currencies.) There is a contradiction between the CMEA practice of inter-government bilateral trade agreements which detail the items to be traded and the Hungarian abolition of production directives to enterprises. The most important factor here is of course the Soviet Union, which accounts for at least half of all trade with socialist countries in the case of every East European country except Romania (see Table 6.7). In fact, any major change in either the administratively planned economy or the method of trading within CMEA would have to be co-ordinated in both spheres because of the degree of mutual dependence which has developed under the present arrangements.

There there is the question of economic priorities. The great attraction of the administratively planned system for the authorities is that it gives them control over the structure of the economy and permits them to ensure that their priorities are followed. (This is why capitalist market economies introduced something close to central planning in wartime.) Whilst in Hungary considerable control over investment, and therefore over the structure of capacity, is retained, the relinquishing of output plans for enterprises inevitably means that the authorities have less influence on the pattern of output and the consumer has more. This is why the level of consumer satisfaction is much higher in Hungary. But many, if not all, CMEA countries' leaders would be unwilling to concede this reduction in their opportunities to enforce their own priorities.

Two specific variants of this general point arise in the case of the Soviet Union: one is the obvious one of the demands of the military sector; the other relates to the development of Siberia. In the long run the natural resources of Siberia represent great economic assets, but the costs of developing them are very large and the process might well be slowed down if more investment decisions were decentralised to the market-guided enterprises. International experience suggests also that regional (and thus nationality-group) inequalities might be aggravated by a more market-orientated investment pattern. The point is not that market economies are unable to cope with this kind of problem but that the Soviet leaders are likely to see moves towards reliance on market pointers as increasing the difficulties. This argument may seem to be weakened by the emphasis that is now being placed by Gorbachev on channelling investment into modernisation of existing industrial plant rather than building completely new capacity because the former offers quicker returns and therefore the prospect of accelerating economic growth sooner. It is true that this is the direction in which market criteria would probably point, if the relevant markets were established, but in what Gorbachev has been saying there is little suggestion that the authorities should delegate their powers of investment allocation to any kind of decentralised, let alone 'spontaneous' or market-determined, decision-making. The question is more one of the length of time horizon which is appropriate for planned investment decisions, and Gorbachev is pressing for a shorter time horizon, at least for the present.

There are also undoubtedly social and political obstacles in the way of far-reaching economic reforms. The most formidable is probably the sheer weight of vested interests in the continuation of the existing system. Administrative central planning requires a large number of bureaucrats, who form a very well-entrenched interest group, and they rightly perceive any serious move towards a market socialist system as a threat to their power and privileges. Their behaviour is one of the factors which causes the traditional system to reassert itself over partial or less far-reaching attempts at reform. They are able to influence the process of deciding what kind of reforms should be introduced and, even more, because any reform has to be introduced by the existing administrative apparatus, without very strong and united political will from the top leadership they are able to frustrate reform at the implementation stage. This is why institutional reform is an important element of a fully successful package, as Hungary learnt. Unity in the political leadership is vital for the adoption and execution of the decisions to make sweeping changes of this kind against the bureaucrats' opposition. If they can secure support from segments of a divided leadership it is unlikely that a consistent and far-reaching reform can succeed.

Nor can it be assumed that outside the economic bureaucracy, in the enterprise themselves, there is no resistance to reform. Managers of course complain about bureaucratic interference in their enterprises; they would undoubtedly like greater freedom of action in some respects, but it does not follow from this that they would all welcome the challenges which a radical change to a market socialist system would present. Some might, but those who have achieved personal success under the existing system have little reason to support a far-reaching set of changes in the rules of the game they have mastered: they must at least harbour some doubts about whether they would be as good at the new game. It is important to remember that no one now of working age in the Soviet Union, and few in the smaller East European countries, have any experience of working under a market economic system (unless, exceptionally, they have acquired it abroad). Managers in these countries therefore have a very different perspective from that of Western observers comparing and contrasting the deficiencies of administrative central planning with (capitalist or mixed) market economies. To such managers, even the Hungarian model would be quite a big step into the unfamiliar. This was less true of Hungary in 1968 because then it had only had central planning for half of an average working life and anybody aged fifty or more had adult memories of the country as a peacetime market economy.

A radical decentralising reform would also be far from sure to be popular with the general public. Because there is dissatisfaction with living standards, it does mean that reforms which should, if successful, lead eventually to improvement in this respect would enjoy instant consumer support. Consumers are also workers, and in certain ways the existing system may be more comfortable than would the brave new world of market socialism. Overmanning, for example, while a reason for low productivity and low real wages, also means that many jobs are less demanding than they would be in a more efficient and productive economy. Problems of redeployment of labour would also arise if reform led to significant structural change in the economy, as in many cases it would need to in order to be fully effective. It is probably true that the majority of people everywhere are innately conservative. This is not to argue that most of the population in CMEA countries are solidly opposed to serious economic reform, let alone that they enthusiastically support the present system. But it is important to remember that a desire for the result of reform is not equivalent to support for the process as it unfolds step by step. The initial impact of some reform measures could easily be unpopular with a public to whom, once again, it would be a move into the unfamiliar. This factor too, therefore, causes leaders to be cautious, even apprehensive, about tackling major reform and is another reason why unity among the leadership itself is an important prerequisite for such reform: without unity, initial unpopularity could be exploited by politicians opposed to the incumbent general secretary and his supporters.

The international climate may also influence the willingness of Soviet and other CMEA leaders to embark on thorough reforms of their economies. Whilst it might be thought that the pressure which deterioration in international relations and higher military spending must place on the economy of the Soviet Union would stimulate reforms by enhancing the urgency of improving productivity, this argument is probably outweighed by the greater priority the leadership would attach to economic control under such circumstances. This is not a topic on which one can cite hard evidence, and discussion is inevitably speculative. The preponderant line of argument, however, seems likely to be that because major economic reform would involve uncertainty, and a tense and unsettled climate of international relations also means greater uncertainty, any leaders would avoid confronting both sets of uncertainties simultaneously. This is reinforced by the point mentioned earlier that in the short run reform would have economic costs; this would militate against undertaking it when pressure of military demands for resources is particularly high. Equally, serious decentralisation of the economy, with abolition of compulsory output targets, would certainy be seen by the leadership as involving an unacceptable loss of the possibility of enforcing its production priorities during a time of heightened international tension.

The conclusion must be that the weight of internal and external political factors which can be expected to make leaders reluctant to embark on a major reform of the economic model very substantially reinforces the economic obstacles previously examined. Although all the European CMEA countries are considered here (except, of course, Hungary), in practice it is probably the attitude of the Soviet leadership that is crucial. Hungary has always been conscious of the suspicion which its reforms might encounter in the Soviet Union, and was probably fortunate that it introduced the NEM originally at the beginning of 1968, when the Soviet leaders appeared more open-minded about economic reform than they had become a year later, after they decided on invasion of Czechoslovakia in order to halt what they saw as unacceptable political change there. The link between such political change and economic reform was not precise, but there clearly was a link, if only because a market socialist reform was part of the Czechoslovak programme. There was a distinct change in the tone of Soviet writing about economic reform from late 1968, with much greater emphasis on the need to preserve the strength of planning and warnings against the chaos caused by too much reliance on the market.[7] It must be questionable whether, had the timetable for the Hungarian reform been one year later, it would have been allowed to proceed at all. In the nature of things the precise extent of Soviet control over other East European countries' policies cannot be ascertained, but it seems doubtful if another country could now follow Hungary's path. This political dependence is reinforced by the difficulties of developing the economic interdependence and integration which are the objectives of the CMEA organisation among countries with widely differing economic models. The CMEA has evolved on the basis of administratively planned economies in the member countries; Hungary can be accommodated because it is small and, when necessary, it modifies its system (by issuing orders for the production of certain goods to meet CMEA objectives). But similar developments in some other member countries would greatly increase the complexity of the CMEA trading system. This topic arises again in chapter 9, but it is worth noting here that this is a additional factor which makes the Soviet position critical for the prospect of far-reaching economic reform in any of the CMEA countries.

Prospects for the future

After the above catalogue of obstacles to the introduction and successful implementation of a far-reaching reform along market socialist lines in Eastern Europe and the USSR, the reader will hardly be surprised if this discussion of the prospects for reform in the future starts by saying that, except perhaps in the long term (about which the prudent observer makes no predictions), it seems very unlikely that such a reform will take place. The only factor one can point to which would suggest this possibility is that many Western and some Eastern economists see such a reform as the only effective solution to the range of economic problems confronting the CMEA countries. Against this are all the difficulties, uncertainties and plain opposition already described. And the CMEA countries' economies, despite their problems, are not in such a critical state—yet—that their leaders are likely to feel compelled to take the risks they would see inherent in such a radical reform.

However, this does not mean that no further attempts at economic reform of any kind are likely in the short or medium term. Quite the reverse. Since his accession to the post of General Secretary in March 1985, Gorbachev has both moved quickly to consolidate his grip on power and has become increasingly outspoken about the failings of the country's economy. Discussion of reform had been reviving during the preceding two years in any case, kindled partly as a reaction to the stagnation of the final Brezhnev years and then by Andropov's reported interest in the subject and his evident determination, until overtaken by ill health, at least to shake up the most lethargic and corrupt corners of the economic bureaucracy. Chernenko was cast more in the Brezhnev mould, but the new General Secretary, who has the great advantage that he can reasonably expect ten years and more in office, has already shown that he attaches great importance to efforts to revitalise the economy and is prepared to tackle the problem vigorously and fearlessly. This was first shown in public criticism of a number of industrial ministers, indications of changes of emphasis in investment policy towards directions offering faster returns, and trenchant declarations of the need to make the economy more efficient.

It is widely thought that a new general 'economic reform' is planned in the Soviet Union, on the basis of the experiment which was started in five industrial ministries in January 1984 and extended in 1985.[8] This amounted to yet another attempt at partial devolution of decision-making and by itself contained nothing to promise any greater success than its predecessors during the 1970s in the Soviet Union and other CMEA countries. The details of the economic mechanism are not, however, the only area in which changes may be introduced. Whilst one cannot expect a radically different economic model to be part of the anticipated Gorbachev reforms, it is premature to exclude the possibility of any improvement in efficiency, at any rate in the shorter term. The possibility of making a basically unchanged system work better is considered at the end of this chapter.

As far as the other countries are concerned, Hungary has been more or less consistently developing its original NEM model since the beginning of the 1980s, after some wavering and a few setbacks during the 1970s. The emphasis recently has been on administrative reform and on gradual alignment of domestic with world market prices.[9] Considerable problems still beset the planning and implementation of investment, the area where most central control was retained and the NEM differs least from the traditional system of central planning. Foreign trade also presents problems, since, as a small economy, the volume of trade must necessarily be large in relation to the country's output because of the dictates of economies of scale. Hungarian planners have consequently been devoting a lot of attention to trying to restructure the economy to be more competitive in Western and Third World markets, so as to mitigate the difficulties of fitting into the CMEA system and of the rising costs of CMEA raw material supplies, and to stimulate efficiency in Hungarian industry. At present the signs are that these various trends of development will continue; Hungary appears to have surmounted the dangers of slipping back into the traditional system and can be expected to pursue its careful and considered policies of building on what has been achieved by the reforms hitherto, despite considerable resistance from some powerful vested interests in the unprofitable industries which must be affected.

Poland presents something of an enigma. During 1981 a reformed economic system was adopted for implementation from January 1982. It was closely modelled on the Hungarian system,[10] and was actually more radical in one respect, namely in providing for a significant measure of workers' participation in management, which is not the case in Hungary. Despite the introduction of martial law in December 1981 the reform officially went ahead. Yet it is clear that a lot of central intervention and direction has continued. The formal position is that the principles that enterprises should be independent, self-financing and self-managing, that prices should be much more flexible (though with some control retained—as in Hungary) and financial discipline much tighter, and that central plan objectives should in the main be realised by the use of indirect instruments, cannot be fully implemented immediately but are regarded as desirable goals which will be reached in due course. In practice a considerable amount of production is still to central and ministerial order, and allocation of scarce materials and intermediate goods in the same way is widespread. The most important and sensitive prices are not free, but many others are decontrolled, though the tax system works to undermine enterprises' interest in earning profits and reducing costs. Evidently the situation at the moment is very mixed and confused, but it has to be said that the experience of the previous attempt at gradual movement to a type of market socialism a decade earlier in Poland does not bode well for the outcome this time. The country's debt situation is obviously going to mean that hard-currency imports remain scarce and must be rationed in some way for the foreseeable future, and this factor is likely to add to the other inherent pressures for retention of a broad measure of central administration, leaving the formal enterprise independence an empty formula.

The other smaller East European countries can probably be expected to follow any lead given by the Soviet Union (though this may not be true of Romania). In particular, the greater emphasis which Gorbachev is placing on economic calculations and rationality in his pronouncements about Soviet investment policy, for example, presage a still tougher attitude on the part of the Soviet Union over supplies of energy and raw materials. This would affect East Germany and Czechoslovakia above all, and could compel them to tackle restructuring of their economies in less material- and energy-intensive directions as well as to concentrate greater attention on using these inputs more efficiently in their existing industries. But any efforts they may make to do this seem more likely to take the form of specific incentives and campaigns for the purpose, within the present economic system, than fundamental changes in system.

The traditional system of administrative central planning, then, looks unlikely to be abandonned in favour of a market socialist model for the foreseeable future in the Soviet Union or Eastern Europe (with just the possible exception of Poland). Does this mean that no improvement at all in their economic performance can be anticipated? In the short and medium term not necessarily so, despite the inherent weaknesses of the traditional system which make its long-run viability doubtful. The evidence of the brief Andropov period in the Soviet Union, although not conclusive, does suggest that the system can be made to operate rather more efficiently than it did in Brezhnev's last years. A general tightening up of work discipline at all levels and in all sectors of the economy, including the economic administration, probably can bring a once-and-for-all gain. A clampdown on abuses and corruption, the diversion of resources for personal benefit, which had evidently become very widespread, can also yield some improvements, partly by making the previously misappropriated resources available for planned purposes again and perhaps also by improving the morale of those unable to benefit from corruption. The point of principle here is that one can in general distinguish more or less efficient administrative bureaucracies and that there are lots of signs that the administrative bureaucracy running the Soviet economy in recent years has been a distinctly inefficient one. Hence there must logically be scope for improvement in this quarter.

Another avenue which could offer substantial gains is the stricter application of criteria of economic rationality to decision-making, most notably to investment decisions. One of the critical weaknesses of central investment planning is that all decisions are exposed to political lobbying by sectoral and regional interests in a way that they are not normally in market economies, at least in the private sector. Gorbachev has talked of the need to concentrate investment where the returns are greatest. If he were able to achieve a significant reduction in the degree to which non-economic pressures influence investment decisions the medium-, though not short-term, benefit could be large. This would be an extremely difficult task, but it is an area where the potential for increasing productivity and growth without a basic systematic change may be great. Even without this element, there is probably scope for some revival of economic growth by the more rigorous concentration of investment in modernisation and re-equipment of existing plants, with less greenfield development, for which Gorbachev has called, though it should be noted that this is by no means a new idea in Soviet economic policy. Efforts to do it have been under way since the 1970s but it proved to be less straightforward than it sounds; one trouble is that old factory buildings are often unsuitable and much of the modernisation budget is swallowed up trying to adapt them. Equipment costs are also inflated when machinery has to be specially made for installation in existing plants. Neither the construction nor the engineering industry in the Soviet Union is geared to renovation work and both try to avoid it.[11]

This illustrates the importance of not overlooking economic policy, as distinct from the economic system,[12] as a reason for good or bad economic performance. Since what is distinctive about the CMEA countries, from the viewpoint of the Western observer, is their economic system, there is often a tendency to concentrate too heavily on this as the source of economic difficulties. There may be opportunities to improve economic performance by policy changes in a number of other areas, and the Soviet leadership is likely to see these as less risky than embarking on major systemic reform. The Soviet economy has not ground to a halt, although it does use resources very inefficiently, resulting in chronic shortages and a low standard of living for its people. At some point, probably quite a long time into the future, its performance may become so bad that the leadership is forced to take the risks of systemic reform. As long as it is unable to convince itself that it is not under constant threat in a hostile and aggressive world, it is likely to go on living with the system it knows, trying to make it work better, to be sure, but insisting on retaining the concentration of economic control in government hands which that system, more than any other, is able to ensure.

Notes

1. For a detailed examination of this, see Philip Hanson, 'Success Indicators Revisited: The July 1979 Soviet Decree on Planning and Management', *Soviet Studies*, vol. 35, no. 1, January 1983, pp. 1–13.
2. A survey of this and of developments in all the European CMEA countries, plus Yugoslavia and Albania, may be found in Alec Nove, Hans-Hermann Höhmann and Gertraud Seidenstecher (eds), *The East European Economies in the 1970s*, London, Butterworth, 1982.
3. See P. T. Wanless, 'Economic Reform in Poland 1973–1979', *Soviet Studies*, vol. 32, no. 1, January 1980, pp. 28–57.
4. For an interesting comparison of Polish and Hungarian experience see P. G. Hare and P. T. Wanless, 'Polish and Hungarian Economic Reforms—A Comparison', *Soviet Studies*, vol. 33, no. 4, October 1981, pp. 491–517.
5. On this see Janusz G. Zielinski, 'On System Remodelling in Poland: A Pragmatic Approach', *Soviet Studies*, vol. 30, no. 1, January 1978, pp. 3–37.
6. See Hare and Wanless, op. cit. and Zielinski, op. cit.
7. Two examples are the unsigned leading article in *Planovoe khozyaistvo* no. 11, 1968, pp. 5 ff. and the article by the senior economist, Academician Khachaturov, in *Vestnik Akademii Nauk SSSR*, no. 12, 1968, pp. 19–26.
8. See, for instance, H.-H. Höhmann, 'Economic Reforms in the USSR: Ways to Raise Efficiency or Blind Alley?' in *The Soviet Union 1984–85*, Westview Press (Praeger, 1986).
9. See Paul G. Hare, 'The Beginning of Institutional Reform in Hungary', *Soviet Studies*, vol. 35, no. 3, July 1983, pp. 313–30.
10. See Stanislaw Gomulka and Jacek Rostowski, 'The Reformed Polish Economic System 1982–1983', *Soviet Studies*, vol. 36, no. 3, July 1984, pp. 386–405.

11. See Boris Rumer, 'Problems of Soviet Investment Policy', in Höhmann, op. cit.
12. For a valuable discussion of this see Antoni Chawluk, 'Economic Policy and Economic Reform', *Soviet Studies*, vol. 26, no. 1, January 1974, pp. 98–119.

9 CMEA trade: problems, prospects and East–West relations

Two views of the problems of the CMEA

The initial account of the operations of the CMEA in Chapter 6 indicated some difficulties which had arisen during the course of its existence and development. It is logically to be expected, however, that what one regards as the major problems facing an organisation will depend on one's view of its objectives, and the CMEA's members are by no means unanimous in their view of the Council's objectives. Whilst in practice there is a spectrum of varying conceptions of how it should develop, rather than just two different views, it is analytically helpful to simplify them into two distinct views, which can be labelled the 'Hungarian' and the 'Soviet', after their principal exponents.

Just as Hungary has been the one CMEA country to introduce reforms to the traditional central planning system which involve elements of the market mechanism, so the 'Hungarian' view of how the CMEA should evolve focuses on the development of financial mechanisms designed to promote more efficient patterns of trade and specialisation by member countries. To the extent that this would mean greater integration of their economies, it would be integration through the unobstructed growth of trade perceived to be mutually advantageous.

From this point of view, the principal problems of the CMEA trading system are its endemic bilateralism, including structural bilateralism, and the difficulty of assessing the real profitability of any trade deal. The restriction of trade which the strong pressure for bilateral balancing involves, and the insistence on matching sales and purchases of 'hard' goods, are serious obstructions to multilateral trade patterns and what could be mutually beneficial specialisation. These problems stem partly from the absence of meaningful prices. What is really meant by the term 'hard' is that the prices of these goods are below the value placed on them by both buyer and seller countries. When all prices are free to move to market-clearing levels, the 'hard' and 'soft' distinction does not exist. In the 'Hungarian' view it would be beneficial to align CMEA prices much more closely with world market prices, and introduce more price flexibility, as indeed is being attempted within Hungary. There has also been a lot of emphasis on development of the financial system of the CMEA, to try to make multilateral trade easier and more attractive. The basic need is to be able to convert a surplus earned on trade with one country into means of payment for goods desired from another country. The fundamental obstacle to this, however, is not the inconvertibility of, say, the zloty into roubles or vice versa, but the inconvertibility of either the zloty or the rouble into Polish or Russian goods. This is a consequence of the internal systems of central planning and allocation of goods in the countries concerned rather than of the CMEA. Fully multilateral trade on anything more than a marginal scale between member countries is not compatible with the prevailing goods inconvertibility within members' domestic economies. (From a different point of view, however, this argument can be reversed and used to advocate integration by means of full supranational planning of the economy of the entire CMEA area as a single country.) For this reason, major progress in the direction of multilateral trade depends on a far-reaching reform of the CMEA members' domestic economic systems in a market-socialist direction, something which looks highly unlikely for the forseeable future.

Another serious problem of the present CMEA system, in the 'Hungarian' view, is the way in which it isolates or insulates the members' economies from developments on the world market. Whilst this may have seemed an advantage during the 1970s, when the Soviet Union supplied oil to the East European countries at prices well below the rapidly rising world level, the benefit has been transitory as the Soviet Union has both refused to continue increasing deliveries after 1980 (and in some cases has subsequently reduced them) and has demanded more and higher-quality goods in exchange, including many goods with a substantial content of Western imports. The terms of trade on which Soviet energy and raw material supplies can be bought have thus deteriorated, and this trend is likely to go further (and in recent years the CMEA five-year moving average price has in any case been catching up with the static or falling world price). More fundamentally, the insulation from world price trends removes stimuli to the technical progress needed to keep abreast of world standards: there has been a much greater reduction in energy consumption in Western than in Eastern Europe during the past decade. Hungarian and other East European economists fear that the CMEA bloc may become locked into an increasingly outdated economic structure and technology which will make it less and less competitive on outside markets. The emergence of the group of Newly Industrialised Countries (NICs) in the world economy in recent years both confirms the basis for these fears and exacerbates the problem for CMEA countries like Hungary which seek to sell more on the world market and to open up their domestic economies to the stimulus of world competition.

The structural stagnation which results from this isolation is further confounded in the case of some of the smaller East European countries by the need to produce some products to satisfy the demands of CMEA partners, especially the Soviet Union, when they would not otherwise devote resources to these industries and products at all. This is symbolised by the formal exception to the general principle of abolition of enterprise output targets under the Hungarian New Economic Mechanism—the case where orders to produce items needed to meet CMEA trade obligations may be issued if no enterprise judges such production sufficiently attractive; in practice this is not now of great importance because most of these exports are attractive to Hungarian enterprises, though it is more questionable how beneficial they are to the national economy. But the criticism is not confined to Hungary, nor necessarily associated with its market-orientated reforms: Czechoslovak economists were making the same complaint as long ago as 1968.

In essence, the negotiations which result in the annual trade agreements between member countries pay comparatively little attention to prices and costs. In tune with the essentially administrative and physical rather than financial nature of central planning, they are basically negotiations to settle how much of which items is to be traded among member countries. Questions of comparative advantage in the economist's sense play little part, and political considerations loom large in the decision-making process. Such a process is clearly totally at odds with the 'Hungarian' conception of how the CMEA countries' economies and their trade should evolve.

From the 'Soviet' viewpoint, on the other hand, the problems and priorities look distinctly different. In this conception, the aim is to move towards more effective integration of the CMEA area through better and more comprehensive co-ordination of the planning and production in each member economy. The sheer relative size of the Soviet economy, as well as the Soviet Union's political and military power, inevitably make it the dominant member in the organisation and there is a history of proposals for and resistance to this kind of specialisation and integration dating back to the Khrushchev years. The smaller countries, in varying degrees, have tended to see such proposals as a threat involving diminution of their sovereignty over their own economies. From the Soviet viewpoint this must look like a tiresome obstacle which has obstructed economically rational specialisation. The 1971 Comprehensive Programme did not immediately lead to any significant movement in this direction, but the more specific programmes initiated in the second half of the 1970s took the crucial step of embracing the planning of production in participating countries as well as just trade among them. Given the nature of the member countries' economic systems, this is a critical distinction, since production does not occur unless it is planned and goods cannot be exported unless produced in sufficient quantity, whatever trade agreements have been signed. This topic arises again in the next section, in connection with the future prospects of the CMEA; for the moment it is sufficient to note that progress towards this kind of integration has been much slower than the Soviet Union probably would have liked.

Another set of problems in Soviet eyes consists of the apparently insatiable demand of Eastern Europe for energy and raw materials from the Soviet Union and the poor quality of many of the goods supplied in exchange. In addition, the nominal improvement in the Soviet Union's terms of trade as a result of the increase in energy prices over the past decade has not been fully enjoyed because Eastern Europe has run a large trade deficit with the Soviet Union, meaning that a part of Soviet exports to the area has been effectively unrequited. From the start of the 1980s the Soviet Union placed a ceiling on its oil deliveries because it needed more for sale to hard-currency markets and has been finding it more difficult to produce enough oil (1984 saw the first ever decrease in annual output). The Soviet Union thus has the same interest as the East European countries themselves do in improvement of their efficiency in resource utilisation, particularly reduction of specific energy and raw materials consumption. But Soviet ideas on how these improvements should be sought are quite different from the Hungarian approach of market stimulation by scarcity-reflecting prices, and focus much more on tighter and more disciplined planning and management of the East European economies. Soviet suspicions of and objections to the more far-reaching reform proposals may well obstruct the improvements for which the Soviet Union itself is also pressing, though the Soviet attitude is of course only one of many barriers to serious reform in Eastern Europe, and is also exploited by elements in those countries who would lose under such reforms and the restructuring of the economy which they would bring about.

The present trend and future prospects

One effect of the differences among CMEA members' conceptions of how the organisation should evolve has been to retard change of any kind. Soviet schemes for greater integration, implicitly leading to if not explicitly proposing supranational planning, date back to Khrushchev, but the actual progress in this direction has been mainly confined to the target programmes of the late 1970s and early 1980s. At various times ambitious proposals for a much greater degree of multilateral settlement through forms of financial convertibility have been put forward by other members, but progress in this direction has been minimal—though this is also attributable to the obstacle presented by goods inconvertibility within member countries' economies, as previously explained. Frequently, therefore, failure to agree on change has resulted in prolongation of the status quo, a phenomenon universally familiar.

Nevertheless, a number of factors have combined to produce an identifiable trend in recent years. This trend is exemplified by the Agreed Plans and the target programmes. It is clear that the Soviet Union has been the prime mover in these programmes and their adoption indicates that the Soviet view of the way forward for the CMEA has gradually gained predominance. To say this is not necessarily to say anything about the relative economic advantages derived from the programmes by the Soviet Union and its partners, though the fact that at least some members would probably have preferred to follow a different path does tell us something about relative weight and influence within the organisation. It is also legitimate to conclude from the fact that the Soviet Union is the proposer of the various schemes that it at any rate sees them as beneficial to itself, particularly if it can use East European dependence to secure the supplies of food products and industrial consumer goods which it is pressing Hungary, for example, to provide in exchange. Essentially what has happened is that the deterioration in the smaller East European countries' terms of trade during the 1970s, and their growing difficulty in selling enough on hard-currency markets, have made them more dependent than ever on the Soviet Union for energy and raw materials. The Soviet Union has thus been able to dictate the terms on which continued supplies will be made available, and these have involved participation in the various joint programmes. Their basic principle is that the East Europeans help to finance the investment and to provide the physical equipment for the development of natural resource deposits in Soviet territory; the production of the equipment is included in their own industrial output plans, and in return they receive supplies from the project when it becomes operational. Whilst not all the programmes are concerned with natural resources, this is the area in which they started and most progress has been made.

The general uncompetitiveness of the CMEA countries on world markets and the substantial levels of debt which some of them accumulated during the 1970s—Poland is the most notorious but not the only case—have also acted to increase their dependence on the Soviet Union by further reducing the possibility of turning to other sources of supply outside the CMEA, for which, of course, payment in convertibility currency would be required. The world recession after 1979 also contributed to the difficulty the East European countries experience in selling outside the Eastern bloc, and it is notable that after 1980 there was a reversal of the trend to reduce the proportion of their trade with socialist countries, particularly on the import side (see Table 6.7), in both Hungary and Romania, the two countries which had achieved the greatest geographical diversification.

The present trend in the trade and development of the CMEA countries is thus for a gradual increase in integration and interdependence achieved through extension of international co-ordination under the aegis of the CMEA to the planning of member countries' production. Because of the political predominance of the Soviet Union, this means in practice some extension of Soviet control over the smaller countries' economic development. It also means trade diversion, reducing the proportion of CMEA members' trade which is conducted with non-members and increasing the proportion within the group above what it would be if the pattern of trade were determined solely by individual economic advantage in a free market. From the point of view of economic theory, such diversion involves a reduction in efficiency and increase in costs. Whilst it is unrealistic to suggest that without the constraint of CMEA membership any of the smaller East European countries could move quickly to completely open trade policies, the present trend represents a shift in the opposite direction. The effect is similar to that of an external tariff, but the level of tariff to which it corresponds cannot be calculated. All one can say is that some advantageous trade opportunities are forgone for the sake of less advantageous ones. The consequent relative isolation from world competition will tend to lock the CMEA countries into backwardness by protecting and preserving their inflexible and inefficient economic systems and structures. This will remain true even if future investment is more carefully attuned to obvious resource endowments than has often been the case in the past.

In fact, of course, with the present price systems in the CMEA countries it is very difficult to know if any particular trade deal is economically advantageous or not. This mirrors the lack of investment criteria in domestic planning. If the ultimate objective is economic efficiency, there is little doubt that profitability at world market prices is the best *general* guide. The objective in the minds of the Hungarian economists advocating integration in the world market is evidently economic efficiency. The objective of the Soviet Union is probably a tightly constrained economic efficiency, subject to the satisfaction of various prior policy goals: one of these is the preservation and extension of its sphere of influence through economic interdependence.

For the foreseeable future the present trend looks set to prevail. But this does not invalidate the analysis which concludes that in the long run such regional production integration and relative isolation will bring about increasing backwardness. Eventually this cost may become unacceptably large, and then fundamental changes in both CMEA trading practices and the member countries' domestic economic systems, which are logically interconnected because of the phenomenon of goods inconvertibility, would have to be introduced. This, however, is to look rather a long way into the future. All the current signs are that under the new and more energetic leadership of Gorbachev the Soviet Union will be both stepping up efforts to make its own existing economic planning system operate more efficiently and increasing pressure on its CMEA partners to raise the quantity and quality of the products they supply and to increase the efficiency with which they use Soviet-supplied raw materials to produce them.

The role of East–West relations

From both the political and the economic points of view the decade of the 1970s witnessed unrealistic hopes of *détente* and growing trade between East and West followed by disillusionment. Politically, Soviet expansion into the Third World, most notably Africa, combined with a continued military build-up, generated a much more hard-headed and suspicious view of Soviet intentions on the part of the major Western powers. Although in Britain, the United States and West Germany this change was associated with a move to the right in domestic politics, it occurred in France, too, where the internal political trend at the time was in the opposite direction. Economically, although there was a marked increase in East–West trade, the more optimistic forecasts of the eary 1970s remained unfulfilled and the share of East–West trade in world trade as a whole is still small. The CMEA countries cannot be said to have become 'full members' of the world economy. From their own point of view, they ended the decade with debts to the West which in some cases represented a serious burden, and it was difficult to identify commensurate benefits. Although some modernisation had been carried out, their overall technological level had not come significantly closer to that of the most advanced countries, and in some cases their capital imports had made them dependent on continuing imports of current supplies because the quality of East European or Soviet imports was too low for satisfactory use in the plants bought from the West.

Poland's debts reached crisis level by 1980, exacerbated of course by the country's political problems, and Romania, Hungary and East Germany also had quite burdensome debt-service ratios by the end of the 1970s. Poland's complicated situation has included rescheduling of its debts, while the other countries have succeeded in containing theirs, and Romania has achieved a notably rapid reduction by means of a drastic austerity programme. The consequences of this experience for future East–West trade are twofold: imports and borrowing from the West will undoubtedly continue to be viewed much more cautiously, as they have been since 1980, and the rapid growth of sales opportunities for some Western industries is not likely to recur. In this sense the lesson of the 1970s borrowing spree has been learnt—in the West as well as the East! But there is little probability of a major contraction of East–West trade either: the East European countries are not in a position to turn their backs on the West and adopt an isolationist approach. They must export to hard-currency markets in order to earn the means to service and repay their debts and they have a continuing need both for certain current supplies and for modern technology which they cannot provide for themselves. The smaller East European countries may be able, and may be forced, to look to Third World countries for a greater proportion of their exports, because in the advanced countries the demand for their manufactures is limited owing to their relatively low quality and technological standards. In the case of the Soviet Union, the main market for the energy and raw materials which earn it the great bulk of its hard currency is Western Europe.

All these factors indicate that the most probable future trend of

East–West trade is one of slow but continuing growth in the longer term. This prediction would only be invalidated if one or more CMEA countries (in particular Poland) were to default on its Western debts. Because default by one member country would have very severe repercussions for the others (witness Hungary's difficulties in 1982 owing to the Polish crisis), this would essentially be a bloc decision, if it happened, and because of the continuing requirements for 'normal' East–West trade such a default is highly unlikely. Despite the weaknesses of the Soviet economy, its position as the world's second largest gold producer and its natural resource endowment mean that it could and no doubt would bale out a CMEA member rather than allow it to default. The price exacted for such support would be much closer Soviet control over the domestic economy of the country concerned, probably accompanied by severe austerity for the consumer.

The general conclusion about the effect of East–West political relations on trade between the two groups of countries which can be drawn from the experience of the 1970s and the first half of the 1980s is that it is quite limited. The period of political *détente* did lead to an acceleration in the growth of trade and perhaps to a clouding of the judgement of Western bankers (though this did not apply only to Eastern Europe). But the scope for trade expansion was and continued to be restricted by the lack of demand in the West for much of what the East has to offer for sale. The Soviet Union was the most successful East–West trader, avoiding excessive borrowing and obtaining significant technological benefits in selected lines of production, and one major reason for this is that its exports to the West are principally primary products. During the 1970s energy prices were rising rapidly, and sales of primary products can be expanded if need be by price competition in a way which does not apply to manufactures. More recently, weak world energy markets have cut back Soviet earning power (and repeated poor harvests mean an additional demand for hard-currency imports). In the uncertain event of East–West relations taking a very marked turn for the better in the second half of the 1980s, it is highly unlikely that any rapid trade expansion would follow. The CMEA countries could not step up their exports much more (they have been trying hard as it is) and neither they nor Western banks would be inclined to embark on a new round of large-scale loans.

Conversely, the slowdown in East–West trade during the first half of the 1980s is wholly explicable by the economic factors described. The relative unimportance of the political climate is confirmed by the course of Soviet–West German trade. The numerous forms of pressure which the Soviet Union tried to apply in its unsuccessful attempt to bully the West German government into not accepting Cruise and Pershing missiles included suggestions that Soviet–West German trade would be curtailed if the missiles were deployed. In fact, entirely predictably, the effect of deployment on the trade has been insignificant, because the Soviet Union needs its trade with West Germany at least as much as West Germany needs its Soviet trade. Where the political element can be more important is in the situation of competition between Western suppliers for a Soviet order; there the Soviet Union may favour a Western country which has been adopting a more 'friendly' political position over one perceived as hostile, if the choice is finely balanced on the basis of economic considerations. The lesson of this is that the West should resist attempts to open gaps in its ranks by means of this and the many other moves made by the Soviet Union with the same aim. In the face of a firm and united political position on the part of the West the evidence is that trade will continue at the level justified by the economic advantages to the participants.

Conclusion

The economic prospects for Comecon can hardly be described as scintillating. But to say this is only to provoke a similar comment about the economic outlook for many other parts of the world, including the Common Market and even the United States. No country or trading group is without its problems; none was unaffected by the rise in oil prices in the 1970s; and none will escape the effects of the current fall.

On the other hand, Comecon has more than its fair share of problems. In general it has an economic system which may have been the most effective for achieving extensive growth but which has mostly been found wanting in the now necessary switch to intensive development. The CMEA needs much more far-reaching systemic change than the EEC. By and large, industrial and agricultural productivity are low, and the service industries are underfunded. Investment and labour are frequently in the wrong place and, because of politically entrenched and ideologically supported vested interests, cannot easily be moved. With certain notable exceptions—not all of them military—technological levels are low, probably below those of Western Europe as well as of the United States and Japan. There are hostile forces in the West anxious to deny Comecon access to their skills and equipment; and Comecon—even more so after the drop in Soviet foreign income from oil—lacks adequate hard currency to buy overseas technological know-how. Finally, however much Comecon members share the same political beliefs, their economic co-operation is essentially half-hearted and not particularly productive.

However, while all this makes trade with the West particularly difficult, it also makes it—at least from the Comecon point of view—all the more necessary. Even the very basic business of paying off Western debts from the 1970s demands a higher level of exports to the West; and at the same time, import substitution designed to save hard currency in fact makes hard currency more difficult to earn through reducing the quality of potential export goods. Improved trade—preferably coupled with some diminution of the debt burden—is high on the CMEA desiderata list.

This does not mean that the CMEA countries want foreign trade on any terms. The Soviet Union has less need than the others, at least proportionately; and the East European members have shown themselves willing, where there is no alternative, to divert more of their exports and imports to within the CMEA market. The new Soviet leaders have also made it clear that they will not be taken advantage of economically, any more than they will politically or militarily. There is also neither law nor custom that makes CMEA salesmen bargain less assiduously or skilfully than their Western counterparts.

They also have things to sell that outsiders have shown themselves willing to buy. Soviet gas is still a precious commodity for the West, no matter what may yet happen to oil. Soviet raw materials, into the extraction of which so much East European investment is being poured, will long continue to be attractive to Western Europe. East European agriculture and its associated food-processing industries have built up a lucrative market in the West and in the Middle East, in spite of EEC restrictions on imports and the opportunistic EEC reaction to the Chernobyl nuclear disaster and its supposed radiation of neighbouring countries' crops. Other industries connected with agriculture—leatherworking for example—and light industries, for instance in the electrical field, are also valued highly outside. Joint enterprise between, say, West Germany and Poland, or Austria and Hungary, are not luxuries invented simply for East European profit.

So even in purely economic terms, and in somewhat difficult times, there are opportunities for the West to trade with Comecon that are advantageous for both sides. And more could be found, particularly where the Soviet Union or the countries of Eastern Europe are markedly skilled or inventive. There is scope, too, for the development of something approaching the special economic zones that have been established in China to bring together Western capital and local labour to export profitably to the Third World.

Admittedly, Comecon does not make trade with itself easy. Essentially non-convertible currencies, centralised buying and selling agencies, and preference for barter or counter-trade deals, all present unusual obstacles to Western businessmen. The shortage of local capital and the long lead-time necessitated by planning add further disincentives, as does a secretive society's antipathy to Western market research. There are, of course, variations from the norm. Reforms in Hungary and latterly in Poland have made many individual enterprises accessible to foreign salesmen; and specific shortages have encouraged ingenuity at all levels within Comecon in facilitating foreign imports. On the other hand, particularly in the Soviet case, there have been periods when political feelings towards the West have reduced what might otherwise have been a natural willingness to trade.

Yet the obstacles, or the diffidence, are by no means all on the one side. The Japanese and the West Germans have shown what persistent and intelligent effort can do to increase their Comecon trade well above that of any other Westerners. By contrast, American concern at the possible transfer of military or militarily-usable technology to the Soviet Union has acted as a brake on East–West trade, and sometimes as a complete barrier. The EEC's quotas and tariffs act against industrial imports from the CMEA, while its Common Agricultural Policy seems set to restrict farming imports for years to come. In addition, what it means by its proposed bilateral agreements with individual CMEA countries is not at all clear; yet in the meantime, its approach has little impact on energy and raw material purchases in the Soviet Union, allows East Germany to operate almost as an EEC member, but discriminates against the saleable products of other East European economies. Nor is the West in general immune to the debilitating effect of political animosity upon commercial intercourse.

Some of Comecon's problems are exogenous. Exceptionally hard winters in Romania and unusually dry summers in Bulgaria are outside Party control. Stalin and his successors did not determine the extent or the quality of arable land in the Soviet Union. Yet collectivisation was not wished upon Stalin by some external deity; nor does some malevolent spirit disrupt his successors' distribution of fertilisers and machinery or delay their collection and storage of crops. As a result of their recent agricultural reforms the Chinese have managed to make 7 per cent of the world's arable land feed 22 per cent of its population and also produce capital sufficient for the development of industry in the countryside. Harnessing peasant acquisitiveness may make bad Marxism; but it raises both productivity and the quality of output. And the Chinese are now experimenting in a similar way with industry—with seeming success. Their foreign trade is also booming, bringing with it hard currency, high technology, and investment from overseas. Though with a very much lighter touch, they preserve their centrally planned economy and maintain both predominantly public ownership and relatively undifferentiated incomes. So even in a Comecon type society there is a scope for indigenous economic improvement and a beneficial growth in international trade. And here a great deal depends upon Gorbachev—how serious he is about achieving an economic breakthrough, and how able he is to translate hi intentions into policies and his policies into reality. For in the last resort, the CMEA goes where the Soviet Union goes, rather than the other way round.

However, the development of international trade that the CMEA as a whole needs, and may in the end want, does not depend solely upon its own attitudes or behaviour. The state of world trade in general, or the terms of trade for the CMEA in particular, are mostly beyond its control, and in that sense they, too, are exogenous factors. Sometimes, however, the attitudes or the behaviour of the outside world towards the CMEA are a response to CMEA action—in directly or indirectly subsidising exports, for example—or indeed to CMEA inaction, as in the case of Soviet delays in giving information about Chernobyl in late April and early May 1986. The converse is also true. But the main point here is that, although the West needs to trade, and although it claims to want to trade with the Soviet Union and Eastern Europe, it does not yet appear to do enough positively to encourage the exchange of goods and services with Comecon.

Economic isolationism appears to be strong in the United States and to be strengthening in the Common Market. In individual Western countries, such as Britain, it is more often a kind of economic myopia, an inability to perceive the desperate need to try to exploit any and every opportunity to expand foreign trade. There is also a tendency, again particularly in Britain, to go for quick returns and to shy away from markets that take a lot of cultivation and are slow to mature. Sometimes the approach of the West is overly political. Yet it is not always unreservedly hostile, whatever the CMEA may think. Western governments have their own opposition forces to counter or assuage, their own interest groups to satisfy, often with conflicting results: EEC farmers saw the fallout from Chernobyl as a means of protecting their own produce from East European competition, whereas grain-growers in the American mid-West saw it as a sudden opportunity to restore their declining fortunes by selling more to the Soviet Union. Sometimes, in so far as there is a common attitude to Comecon, it is one of imagined superiority or simple indifference. But there is frequently a direct conflict between the political and the economic attitudes of the West. The loans of the 1970s were always meant to be recovered with profit. But there was at least a suggestion of a political hope that they would generally assist the process of economic reform and help loosen excessively close East European–Soviet ties. Now, however, when the interest is hard enough to secure, let alone the principal, the emphasis being placed upon probity in economic matters is having the political effect of driving Eastern Europe even closer to the Soviet Union—though perhaps that is also intended as a political lesson.

To put it another way, politics cannot be eliminated from the attitudes or the behaviour of Western countries any more than it can be from those of Comecon. Yet it frequently clouds approaches to Comecon that aim to develop economic relations and that might have advantageous political consequences. Hanging over every other issue, of course, is the wider question of East–West relations with its military ramifications and economic repercussions. That is not the subject of this book, although there is no point in running away from the fact that a new or a renewed arms race would further distort the economic development of Comecon and make a growth in its more conventional foreign trade highly unlikely. But conversely, the lure of an improved foreign trade could make a contribution to reducing East–West tension. However that may be, there are certainly opportunities now for the West to reinvigorate its trade with Comecon; but they need to be sought and to be worked at; and they also need to be used with more than a modicum of realism, clear-headedness and sincerity. And on balance it would seem wiser to encourage whatever are the more promising trends to be perceived from time to time within the CMEA.

The economic development of Comecon and the international trade inextricably connected with it are likely to benefit not only the Soviet Union and its East European neighbours but the West as well. But for neither party will the road be easy.

Select bibliography

S. Ausch, *Theory and Practice of CMEA Cooperation*, Budapest, Akadémiai kiadó, 1972.
J. Berliner, *The Innovation Decision in Soviet Industry*, Cambridge, MA, and London, MIT Press, 1976.
O. T. Bogomolov, *Strany sotsialisma v mezhdunarodnom razdelenii truda*, Moscow, Nauka, 1986.
A. Boltho, *Foreign Trade Criteria in Socialist Economies*, London, Cambridge University Press, 1971.
J. M. P. van Brabant, *Bilateralism and Structural Bilateralism in intra-CMEA Trade*, Rotterdam, Rotterdam University Press, 1973.
J. M. P. van Brabant, *Essays on Planning, Trade and Integration in Eastern Europe*, Rotterdam, Rotterdam University Press, 1974.
J. M. P. van Brabant, *Socialist Economic Integration: Aspects of Contemporary Economic Problems in Eastern Europe*, Cambridge, Cambridge University Press, 1980.
A. Brown and M. Kaser (eds), *Soviet Policy for the 1980s*, London, Macmillan, 1982.
J. F. Brown, *Bulgaria Under Communist Rule*, London, Pall Mall, 1970.
R. W. Campbell, *Soviet-type Economies*, London, Macmillan, 1974.
D. Childs, *The GDR: Moscow's German Ally*, London, Allen & Unwin, 1983.
D. Childs (ed.), *Hönecker's Germany*, London, Allen & Unwin, 1985.
O. A. Chukanov (ed.), *Khozjaistvenni mekhanizm v stranakh-khlenakh SEV*, Moscow, Politizdat, 1984.
C. Coker, *The Soviet Union, Eastern Europe and the New International Economic Order*, Washington DC, Sage, 1984.
W. S. Conyngham, *The Modernisation of Soviet Industrial Management*, Cambridge, Cambridge University Press, 1982.
K. Dawisha and P. Hanson (eds), *Soviet-East European Dilemmas*, London, Heinemann, 1981.
D. Dyker, *The Soviet Economy*, Granada Publishing, Crosby Lockwood Staples, 1976.
S. Fischer-Galati, *Eastern Europe in the 1980s*, Boulder, Colorado University Press, 1981.
S. Fischer-Galati (ed.), *The Communist Parties of Eastern Europe*, New York, Columbia University Press, 1979.
O. Gado (ed.), *Reform of the Economic Mechanism in Hungary: Developments, 1968–81*, Budapest, Akadémiai kiadó, 1982.
M. I. Goldman, *USSR in Crisis: The Failure of an Economic System*, New York and London, Norton, 1983.

SELECT BIBLIOGRAPHY 171

D. Granick, *Enterprise Guidance in Eastern Europe*, Princeton, Princeton University Press, 1976.
P. Gregory and R. C. Stuart, *Soviet Economic Structure and Performance*, New York, Harper and Row, 1974 and 1981.
P. Hanson, *Trade and Technology in Soviet-Western Relations*, London, Macmillan, 1981.
E. A. Hewett, *Foreign Trade Prices in CMEA*, London, Cambridge University Press, 1974.
H- H. Höhmann, M. Kaser and K. Thalheim (eds), *The New Economic Systems of Eastern Europe*, London, C. Hurst, 1975.
F. Holzman, *Foreign Trade Under Central Planning*, Cambridge, MA, Harvard University Press, 1974.
F. Holzman, *International Trade Under Communism*, London, Macmillan, 1976.
M. C. Kaser, *Comecon*, London, Oxford University Press, 1967.
J. Kornai, *The Economics of Shortage*, Dordrecht, North Holland Publishing, 1980.
A. Köves, *The CMEA Countries in the World Economy: Turning Inwards or Turning Outwards*, Budapest, Akadémiai kiadó, 1985.
B. Kovrig, *Communism in Hungary from Kun to Kádár*, Stanford, CA, Hoover Institute Press, 1979.
F. Kozyma, *Economic Integration and Economic Strategy*, Budapest, Akadémiai kiadó, 1982.
J. A. Kuhlman, *The Foreign Policies of Eastern Europe: Domestic and International Determinants*, Leyden, A. W. Sijthoff, 1978.
J. R. Lampe, *The Bulgarian Economy in the Twentieth Century*, Beckenham, Croom Helm, 1985.
M. McCauley and S. Carter (eds), *Leadership and Succession in the Soviet Union, Eastern Europe and China*, London, Macmillan, 1986.
P. Marer and J. M. Montias, *East European Integration and East-West Trade*, Bloomington, Indiana University Press, 1980.
O. A. Narkiewicz, *Eastern Europe 1968-84*, Beckenham, Croom Helm, 1986.
A. Nove, *The Soviet Economic System*, London, Allen & Unwin, 1977.
A. Nove, H-H. Höhmann and G. Seidenstecher (eds), *The East European Economies in the 1970s*, London, Butterworth, 1982.
K. Pécsi, *The Future of Socialist Economic Integration*, New York, Sharpe, 1981.
F. Pryor, *The Communist Foreign Trade System*, London, Allen & Unwin, 1963.
T. Rakowska-Harmstone and A. Gyorgy (eds), *Communism in Eastern Europe* Bloomington, Indiana University Press, 1979.
C. Saunders (ed.), *East-West Trade and Finance in the World Economy*, London, Macmillan, 1985.
G. Schiavone, *The Institutions of Comecon*, London, Macmillan, 1981.
G. Schiavone (ed.), *East-West Relations: Prospects for the 1980s*, London, Macmillan, 1982.
V. I. Sedov, *The Socialist Community at a New Stage*, Moscow, Progress Publishers, 1979.

M. Shafir, *Romania: Politics, Economics and Society*, London, Frances Pinter, 1985.

M. D. Simon and R. E. Kanet (eds), *Background to Crisis: Policy and Politics in Gierek's Poland*, Boulder, CO, Westview Press, 1981.

A. H. Smith, *The Planned Economies of Eastern Europe*, Beckenham, Croom Helm, 1983.

G. A. Smith, *Soviet Foreign Trade: Organisation, Operations and Policy, 1918–1971*, London, Pall Mall, 1973.

M. Simai and K. Garam (eds), *Economic Integration: Concepts, Theories, Problems*, Budapest, Akadémiai kiadó, 1977.

V. Sobell, *The Red Market: Industrial Cooperation and Specialisation in Comecon*, Aldershot, Gower, 1984.

M. J. Sodaro and S. L. Wolchick (eds), *Foreign and Domestic Policy in Eastern Europe in the 1980s*, London, Macmillan, 1983.

R. Szawlowski, *The System of the International Organisations of the Communist Countries*, Leyden and Reading, MA, Sijthoff and Nordhoff, 1976.

S. M. Terry (ed.), *Soviet Policy in Eastern Europe*, New Haven, CT, Yale University Press, 1984.

R. L. Tökes (ed.), *Opposition in Eastern Europe*, London, Macmillan, 1979.

F. Triska and C. Gati (eds), *Blue-collar Workers in Eastern Europe*, London, Allen & Unwin, 1981.

Y. T. Usenko, *The Multilateral Economic Cooperation of Socialist States*, Moscow, Progress Publishers, 1977.

I. Vadja, *Foreign Trade in a Socialist Economy*, London, Cambridge University Press, 1971.

W. V. Wallace, *Czechoslovakia*, London, Westview, 1977.

P. J. D. Wiles, *Communist International Economics*, Oxford, Basil Blackwell, 1966.

P. J. D. Wiles (ed.), *The New Communist Third World*, Beckenham, Croom Helm, 1981.

J. Wilczynski, *The Economics and Politics of East–West Trade*, London, Macmillan,, 1972.

J. Wilczynski, *Technology in Comecon: Acceleration of Technology Progress Through Economic Planning and the Market*, London, Macmillan, 1974.

J. Woodall (ed), *Policy and Politics in Contemporary Poland: Reform, Failure and Crisis*, London, Frances Pinter, 1982.

Index

Abrasimov, Pyotr 47
Afghanistan 20, 34, 43, 44, 101
Africa 44
agreed plans 99, 105, 159
agriculture 6–7, 12, 17–18, 25–6, 28, 51–2, 55, 58, 64, 72–4, 76, 78, 87, 92, 96, 118–19, 128–3, 166–7
Albania 73
Andropov, Yuri 16, 46, 52, 97, 138, 149, 151
Angola 99
Asia 44
Austria 21, 166
autarky 3, 49, 56, 74, 102

bargaining 137
Berlin 30, 31
bilateralism 2, 49, 53, 55, 67, 73–4, 103–4, 155
Böhlen ethylene complex 70
Brezhnev, Leonid ix, 7, 10–12, 15–16, 20, 27, 41–5, 48, 51–2, 58, 62, 67, 71, 97, 99, 149, 151
Brezhnev Doctrine 8, 38, 41
Britain 36, 168
Budapest 1956 5–6, 12, 23
Bulgaria vii, ix, 3, 7–9, 27–30, 35, 37, 47, 78–80, 99, 116, 128, 167
Bulgarian Communist Party 27–8, 57, 63
Bureaucracy 3, 19, 23, 36, 58, 64, 69, 74, 145–6

Castro, Fidel 49, 99
Ceauşescu, Nicolai 25–7, 36–7, 47, 63, 93
central planning 3, 19, 32, 35, 42–3, 48, 52–3, 55, 58, 64, 70, 73, 79–80, 97, 123–5, 131, 133–5, 141, 152, 157, 167
Charter 77 21
Chernenko, Konstantin 16, 46, 47, 49, 52, 149
Chernobyl 166, 168
China ix, 6–8, 26–7, 34–5, 37, 43–5, 47–8, 64–5, 97–8, 100, 132, 166–7

CMEA *see* Council for Mutal Economic Assistance
Comecon *see* Council for Mutual Economic Assistance
Cominform 1–2, 4, 5
Common Market *see* European Economic Community
comparative advantage 157
Complex/Comprehensive Programme 8–10, 69, 106, 158
comprehensive programme of scientific and technological progress ix, 49, 59
consensus 20
consumption and living standards 16–17, 21, 23–4, 28, 31, 33, 55, 58, 64, 83, 85, 90, 92, 94, 125, 126
convertible currencies x, 10, 49, 69, 74, 104, 121, 128, 166
co-ordination ix, 1, 6, 9–10, 42, 49, 68–70, 73
Council for Mutual Economic Assistance *passim*
  banks 100, 103
  committees 50, 69
  comparison with European economic Community 72–7
  council sessions 4, 7, 48–9, 51, 67–72, 75–6
  Executive Committee 75
  Moscow headquarters 1, 10, 69
Cruise and Pershing missiles 22, 30, 37, 43–4, 46–7
Csaba, László 121, 123
Cuba vii, 10, 73, 98–100
Cuban missile crisis 43
Czechoslovakia vii, ix, 1, 3, 5, 8–12, 20–3, 25, 30, 33, 35–8, 42, 44–6, 49, 80–3, 100, 103, 117, 134, 136, 148, 151
Czechoslovak Communist Party 22, 43, 57, 63

Danube power scheme 33
debts/loans 10–1, 21, 23, 32–3, 36, 47, 49, 57, 79, 82, 84, 87, 89, 93, 95,

INDEX

debts/loans (*cont.*)
   127–8, 140, 151, 160, 162–3, 165, 168–9
détente 7, 10, 37, 50, 57, 65
Devna soda factory 70
Druzhba oil pipeline 70
Dubček, Alexander 20, 23

East Germany vii, ix, 7–9, 30–3, 37–8, 43, 46–7, 49–50, 53, 83–5, 99, 103, 117, 119, 125, 134, 136, 151, 162, 167
East German Socialist Unity Party 31–2, 57, 63
East–West relations xi, 34, 47, 56, 65, 161–4, 169
EEC *see* European Economic Community
Employment, growth of 92, 117
  problems with 88–9
European Coal and Steel Community 74
European Economic Community vii–x, 1, 7, 10, 26–7, 31, 34, 38, 44, 46, 49–50, 56, 64, 65, 72–7, 165, 167–8
  Commission 74–5
  comparison with Council for Mutual Economic Assistance 72–7
  Council (of Ministers) 74–6
  Parliament 74, 76
experiments, economic 142

Federal German Republic *see* West Germany
Figurnov, E. 51–3
financial mechanisms 103, 155–6, 159
France 36, 74

Gdansk 45
Geneva summit 47–8
German Basic Treaty 31–2
German Democratic Republic *see* East Germany
Gierek, Edward 11, 45, 139
Glemp, Archbishop 18, 37
Gomulka, Wladyslaw 45, 138–9
Gorbachev, Mikhail viii–xi, 16–17, 21–3, 27–8, 32–3, 35–6, 46–8, 52–3, 56–65, 72, 77, 97, 138, 143, 145, 149, 151–2, 161, 167
grain imports 95, 163, 168
Greece 37
Gromyko, Andrei 47
growth, decline in 52, 78, 81–2, 116–19, 125–7

hard goods 103–4, 155
Honecker, Erich 30–1, 33, 36–7, 46–7, 63
Hungary vii, x, 2, 5, 8, 10, 12, 18, 22–6, 30, 32–3, 35, 37–8, 43–4, 46, 48–9, 53, 56, 60, 63, 85–8, 91, 103–4, 106, 119, 123, 125, 128, 132, 134, 136, 144, 148, 150, 155, 157, 160, 162, 166
Hungarian Socialist Workers Party 24–5, 37
Husák, Gustav 22–3, 46, 57, 63

IMF *see* International Monetary Fund
incentives 94, 96, 123–5, 135
inconvertibility 103, 156, 161
inflation 140
integration viii–ix, 1–2, 4, 8–9, 16, 32, 35–6, 42, 48–50, 53, 64, 69–71, 76, 105–6, 155–7
International Development Association 35
International Finance Corporation 35
International Monetary Fund 23–4, 35, 100
investment 19, 21, 36, 53, 70–1, 74, 84, 88–90, 92, 95, 116–17, 136, 144–5, 150–1, 165–6
  criteria 88, 152, 161
Iran 21
Israel 37
Italy 36–7

Japan ix–x, 34, 165, 167
Jaruzelski, Wojciech 17–20, 36–7, 45–6, 60, 63
joint financing 96, 104, 105, 159–60

Kádár, János 23, 25, 35–6, 46, 63
Katowice iron and steel combine 70
Khrushchev, Nikita ix, 2–7, 15, 26, 42, 45, 68, 97–8, 105, 157, 159
Kohl, Helmut 46
Kommunist 47
Korean War 3
Kosygin, Alexei 7, 135
Kulikov, Viktor 45

Labour supply 81, 83, 87, 93–5, 117, 119
LDCs *see* less developed countries
less developed countries 92
Libya 21
Los Angeles Olympics 29

management 124, 132, 135, 146, 150
Manevich, E. 120
manning levels 120, 125, 146–7
Marshall Plan 1
Middle East 29, 44, 166
Military expenditure 34, 51, 53, 126, 145
Mir power grid 70
Mongolia vii, 10, 97–8, 99
Moscow Declaration 6
Mozambique 100–1

NATO see North Atlantic Treaty Organization
natural resources vi, 3, 71, 96, 128, 159–60, 166
Newly Industrialised Countries 127, 156
NICS see Newly Industrialised Countries
North Atlantic Treaty Organization viii, 34, 44, 47, 61, 76
Nowa Huta 19

Obzina, Jaromír 21
oil/energy x, 10–11, 21, 23, 26–9, 32–3, 36, 49–50, 52–3, 64, 71, 74–5, 95, 120–1, 158, 165–7
Orenburg pipeline 105

Patriotic Committee for National Rebirth 20
People's Republic of China see China
Poland viii–ix, 1, 5, 7–9, 11–13, 16–23, 25–6, 28, 33, 36–7, 41, 43–6, 49, 88–91, 103, 116–17, 120–1, 125, 127, 130–2, 134, 138–40, 150–1, 160, 162, 166
policy, economic 152–3
Polish United Workers Party 17, 45, 57, 63
Pope John Paul II 45–6
Popiełusko, Father 18, 46
Prague 1968 20–1, 23
Pravda 47–9
prices 84, 89, 93, 95, 97, 99, 120, 123–5, 135–6, 139, 141–3, 150, 155, 161
 of new products 140, 143
 retail 88–90, 126, 132
priorities 144–5
productivity viii, 7, 19, 52, 58, 81, 117, 119–23, 152, 165

Reagan, Ronald 48
reforms, failure of 141–3
 'First Wave' of 7–9, 23, 28, 35, 134–7

obstacles to xi, 8, 23–5, 28, 36, 48, 143–8, 158
religion in Eastern Europe 13, 18, 20, 22, 26, 28, 30, 37, 43–4
Romania vii, 3, 6–9, 25–30, 32, 34–8, 44, 49–50, 70, 91–4, 97, 106, 116, 120, 134, 138, 144, 160, 162, 167
Romanian Workers Party 25–7
Ryzhkov, Nikolai 51, 53, 56, 63

SALT see Strategic Arms Limitations Treaty
SDI see Strategic Defence Initiative
Sellers' market 124, 140, 143
Shirayev, Yuri 51
Shishkov, Yuri 71
Siberia 145
Šik, Ota 135
Sindermann, Horst 37
Socialist 'Commonwealth'/'Community' 8, 18, 22, 27, 30, 42, 51, 59, 61–2
Socialist market economy 7–8, 18, 22–3, 25, 35–6, 48, 58, 136, 139, 140, 143, 146
Solidarity 1, 11–13, 17–20, 45, 63
Soviet Union vii–x, 1–8, 10–13, 16–17, 19–22, 27–38, 41–6, 48–52, 55–9, 62, 64–5, 70–3, 76, 79, 82–3, 94–7, 99–100, 102, 116–17, 121, 124–5, 130, 133, 135, 137–8, 149, 157–9, 166–9
Soviet Communist Party 42, 44, 58–64, 67
 Twentieth congress 5
 Twenty-sixth congress 15–16, 35, 41, 45, 49–51
 Twenty-seventh congress ix, 16, 35, 53, 55–63, 72
Soyuz gas pipeline 70
SS-22 missiles 22, 30, 33, 37, 47
Stalin, Josef viii–ix, 1–6, 15, 42, 130, 167
Stalinists/Stalinism 23, 26
Statistics, accuracy of 86, 116
Strategic Arms Limitations Treaty 10
Strategic Defence Initiative 47, 52
Sweden 37
Szürös Mátyás 37

Target Programmes 9, 10, 70, 99, 105, 159
Technology, new viii, x, 51–3, 55, 58, 63–4, 76–7, 120, 123–5, 175
 imports of x, 7, 10, 11, 31, 50, 64, 127–8, 165, 167

terms of trade 81, 87, 90, 121, 126–7, 156, 158–9
Tikhonov, Nikolai 16, 48–9, 51
trade diversion 160

Ulbricht, Walter 30
United Nations 44
United States vii–x, 8, 10, 34, 43–4, 46–7, 52, 56, 65, 165, 167–8
Ust-Ilmisk cellulose plant 70

Vietnam vii, 10, 73, 95, 99–100
Vintrova, R. 121

Wałesa, Lech 17–18, 45
Warsaw Treaty Organization viii, 12, 27, 34, 36–7, 44–5, 47–8, 53, 61, 74, 76
    Political Consultative Committee 11, 34, 47
    Foreign Policy Committee 11
West Germany 10, 21, 30–3, 37, 46–7, 163–4, 166–7
Winiecki, Jan 121
World Bank 35, 100
World Communist Conference 8
Wrocław 19
WTO *see* Warsaw Treaty Organization

Yalta anniversary 46
Yermakovo ferro-alloy 70
Yugoslavia 5, 12, 37, 44, 136
    Macedonia 29

Zamyatin, Leonid 49
Zhivkov, Todar 27–9, 37, 46, 63
Zhivkova, Lyudmila 28